GRAVITATIONAL SYSTEMS OF GROUNDWATER FLOW

Theory, Evaluation, Utilization

Groundwater of meteoric origin permeates the upper parts of the Earth's crust in spatially organized flow systems down to several kilometres. Since the discovery of the flow-system concept in the 1960s, hydrogeology's basic paradigm has shifted from confined flow in aquifers to cross-formational flow in drainage basins. Consequently, groundwater has been recognised as a fundamental geologic agent, generating and modifying natural processes and phenomena of scientific, practical and economic interest.

This book is the first to present an extensive and illustrated overview of the history, principles, study methods, practical applications and natural effects of gravity-driven groundwater flow. Its user-friendly presentation requires no advanced background in mathematics, with the necessary mathematics being explained in full, and the physical meaning of the equations emphasized. The author highlights significant inter-relationships between the broad range of seemingly disparate processes and systems, demonstrating how these can be traced to a common root cause involving gravity-driven groundwater flow. Examples are used to illustrate practical applications in areas as diverse as hydrogeology, land-use planning, environment protection, wetland ecology, agriculture, forestry, geotechnical engineering, nuclear-waste disposal, mineral and petroleum exploration, and geothermal heat flow.

Written by one of the founding fathers of modern hydrogeology, and including an extensive glossary to aid students and researchers from a variety of disciplines, this book is a key reference for researchers, consultants and advanced students of hydrogeology and reservoir engineering.

JÓZSEF TÓTH began his study of geophysics in Hungary in the early 1950s, but moved to the university of Utrecht in the Netherlands in 1956 following the Hungarian revolution. He later emigrated to Canada, where he joined the Alberta Research Council in 1960. He shifted the paradigm of strata-bound groundwater flow in drainage basins to cross-formational water movement by two ground-breaking papers in 1962 and 1963 before defending his PhD thesis in Utrecht in 1965, and has contributed fundamental concepts and observations to the role of groundwater as a geologic agent. He joined the University of Alberta in Canada as a sessional instructor in 1966 and as a full-time faculty member in 1980. He currently holds the positions of Professor Emeritus at the University of Alberta, and Titular Professor at the Eötvös Loránd University in Budapest, Hungary.

Professor Tóth has received many awards for his work in hydrogeology, including: the first O. E. Meinzer Award from the Geological Society of America in 1965; the 1999 President's Award from the International Association of Hydrogeologists (IAH); the 2002 Prix R. N. Farvolden Award from the Hydrogeology Division of the Canadian Geotechnical Society; the 2003 M. King Hubbert Science Award of the National Ground Water Association (NGWA) of the USA; and the 2004 C.V. Theis Award of the American Institute of Hydrology.

GRAVITATIONAL SYSTEMS OF GROUNDWATER FLOW

Theory, Evaluation, Utilization

JÓZSEF TÓTH

University of Alberta, Canada; Eötvös Loránd University, Hungary

CAMBRIDGE
UNIVERSITY PRESS

University Printing House, Cambridge CB2 8BS, United Kingdom

One Liberty Plaza, 20th Floor, New York, NY 10006, USA

477 Williamstown Road, Port Melbourne, VIC 3207, Australia

314-321, 3rd Floor, Plot 3, Splendor Forum, Jasola District Centre, New Delhi - 110025, India

79 Anson Road, #06-04/06, Singapore 079906

Cambridge University Press is part of the University of Cambridge.

It furthers the University's mission by disseminating knowledge in the pursuit of education, learning and research at the highest international levels of excellence.

www.cambridge.org
Information on this title: www.cambridge.org/9781108460545

© J. Tóth 2009

First published 2009
First paperback edition 2018

A catalogue record for this publication is available from the British Library

ISBN 978-0-521-88638-3 Hardback
ISBN 978-1-108-46054-5 Paperback

To my wife Erzsike, who has patiently endured the many lonely days, weeks and months that I have devoted to my hobby

Contents

Preface

This monograph is intended to present a personal perception of the birth, evolution and consequences of a single geological concept: the gravitational systems of groundwater flow. The concept seems to have been instrumental in redefining the scope of a single-issue water-supply problem into the many-faceted earth science sub-discipline of modern hydrogeology. It has shifted the paradigm of aquifer-bound groundwater flow to cross-formational water movement in hydraulically continuous drainage basins.

This view was corroborated recently by approximately twenty-five papers presented at the Annual Meeting of the Geological Society of America, Denver, October 28–31, 2007, in the two sessions of Topic 34, 'Regional Groundwater Flow: ...' The papers demonstrated a still lively interest in the topic 45 years after the first publication of the concept (Tóth, 1962a, 1962b), a still broadening scope of research, and an increasing variety of practical applications, as exemplified by Glaser and Siegel (2007), Gleeson and Manning (2007), Mádl-Szőnyi, (2007), Otto (2007), Rudolf and Ferguson (2007) and Winter (2007).

My perception has evolved from my own research, practical experience and literature studies in hydrogeology over 47 years (Tóth, 2002, 2005, 2007). It is presented here as a distilled summary of the relevant parts of my earlier publications, lectures and courses. I hope to summarize the results and the consequences of that work in the form of a consistent, coherent and all-round story. Illustrative case studies and case histories have been taken from my field and theoretical work as well as from published literature.

The subject matter is the basinal-scale systems of gravity-driven natural groundwater flow. It is recognized, however, that the term 'basinal scale' is relative and the question is addressed explicitly (Section 3.1.3). Conceptually, the discussion is focused on three main aspects of the topic: (i) the mathematically formulated theory of the formation, evolution and controlling factors of gravity-driven flow

systems; (ii) the methods of and approaches to the practical evaluation and portrayal of those flow systems; (iii) the hydrological, hydrogeochemical, geothermal,
geotechnical, mineralogical, pedological, botanical and ecological factors, i.e. the
natural consequences and manifestations of flow systems.

Since the publications of Domenico (1972) and Freeze and Cherry (1979), most
monographs and textbooks on hydrogeology discuss certain aspects of gravitational
flow systems of regional scales. In the present treatise I try to give a comprehensive
overview of all the principal aspects of the topic.

Hydrogeology is not treated generally at the introductory level in the book. Only
basic concepts directly relevant to the subject matter are defined and explained
explicitly. Nevertheless, I have noticed over the years that some concepts and
parameters, particularly those that are novel or not generally discussed in basic texts,
are notoriously difficult for some students to grasp. In order, therefore, to ensure
a thorough understanding of the intended subject I found it necessary to go into
almost elementary detail of the elucidation of such questions. In these explanations
I place the emphasis on the physical content rather than on mathematical pedantry.
In general, owing chiefly to the rapidly increasing popularity of hydrogeology both
in academia and practice, a growing number of professionals from a wide variety
of disciplines probably already possess the required basic knowledge in hydrogeology to be interested in the book. I would thus expect the primary readership to be
graduate students, researchers and consultants in the various disciplines mentioned
above. However, undergraduate students in hydrogeology, land-use planners and
administrators in water and natural resources may also have an interest in some
aspects of the topics presented.

A piece of work like this can never be considered to be the product of one single
individual. This book is no exception. I would, therefore, like to acknowledge the
contribution of the many friends, colleagues and coworkers who have broadened
my knowledge, questioned, tested, or complemented my ideas, or warned of possible dead-ends through extended collaboration, chance discussions, private and
published debates and many other means. There are too many such people to name
them individually. Nevertheless, I cannot leave unmentioned my former Graduate
Students at the University of Alberta, Edmonton, Canada, an ever challenging and
rewarding group. I am sure you, my friends, will recognize your contributions sprin-
kled all over the following pages. I must also express my gratitude, while retaining
responsibility for the contents, to Dr O. Batelaan and Dr Zijl for kindly reviewing
and advising on the sections developed from their papers, Dr E. Eberhardt and
Dr G. D. Lazear, for providing me with the originals of some diagrams from their
papers that I discuss, and Dr D. Hansen who, without knowing me personally, spon-
taneously offered to have the derivations in Appendices A and B typed from my
then forty year old hand written script. Last, but by no means least, I am indebted

to the Department of Physical and Applied Geology, Eötvös Loránd University, Budapest, Hungary, for giving me a comfortable and warm working environment for writing this book.

As a participant in this small but personally rewarding part of scientific history it has given me a life-long pleasure to be part of the process.

József Tóth
Budapest, 2008

1

Introduction

1.1 The subject matter: definition, history, study methods

Gravity-driven regional groundwater flow is induced by elevation differences in the water table and its pattern is self-organized into hierarchical sets of flow systems. Tóth (1963, p. 4806) defined a groundwater flow system as '*a set of flow lines in which any two flow lines adjacent at one point of the flow region remain adjacent through the whole region; they can be intercepted anywhere by an uninterrupted surface across which flow takes place in one direction only.*' While flow is generated by the relief of the water table, its patterns are modified by heterogeneities in the rock framework's permeability.

Topographic effects are ubiquitous and may cause water to move at depths of several kilometres beneath the Earth's terrestrial areas. Most of people's needs for subsurface water are met with water obtained from this depth range. However, in addition to satisfying this need, gravity-driven groundwater also generates and affects a wide variety of economically important natural processes at or below the land surface. It is of both economic and environmental importance, therefore, to understand the properties, controlling factors, effects and manifestations of this type of flow, as well as to develop methods and techniques for its study and possible modification. Furthermore, because of the relatively easily accessible depths, known and measurable controlling factors, observable natural effects and manifestations, in short, unique tractability, the study of gravity-driven groundwater flow is instructive and useful in the understanding and exploitation of groundwater motion generated by other sources of driving forces, such as differences in dissolved salt contents, thermal convection, sedimentary compaction and tectonic compression.

Owing to their practical relevance and scientific nature, the questions of driving forces, spatial patterns and controlling factors of natural groundwater flow have long interested hydrologists, hydrogeologists and, more recently, Earth scientists in general. Munn's hydraulic theory of oil and gas migration is one of the many

1

Introduction

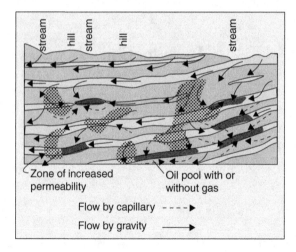

Fig. 1.1 Groundwater flow and hydrocarbon accumulation conceptualized by Munn (1909; modified from Fig. 77, p. 526).

possible examples that illustrate the point. Munn (1909) envisaged meteoric water to descend from the land surface across beds of sandstone and shale and, driven by capillary forces, to push ahead particles of oil and gas dispersed in these beds (Fig. 1.1).

Permeability differences in the rock would cause different parts of the fluid front to advance at different rates 'which would finally result in zones of conflicting currents of water between which the bodies of oil and gas would be trapped and held' (Munn, 1909, Figs. 77–79; p. 525). The idea of conflicting currents of groundwater appears to be a realistic mechanism for entrapment (its application to petroleum exploration based on the theory of gravity-driven flow systems will be shown in Section 5.5). However, sites where such conditions might occur cannot be identified in practice from Munn's concept because the relations between flow directions and the factors controlling them are not specified.

Perhaps the earliest published conceptualization of hierarchically distributed groundwater flow systems is reproduced by Fourmarier in his 'Hydrogéologie' (1939, Fig. 43, p. 87, from D'Andrimont, 1906). Figure 1.2 shows a major water divide with a sub-basin to the left from its crest. From both sides, the sub-basin attracts two, what we call today *local*, groundwater flow-systems. The local systems are superimposed on a larger system that originates on the principal divide and moves towards the main valley of the watershed. A similar notion seems to be reflected by two tiny flow systems leading to a saline 'Discharge area in sidehill valley' on the right-hand side flank of Meyboom's (1962, Fig. 2, not reproduced here) 'Prairie Profile'. But the minuteness of the feature suggests a lack of

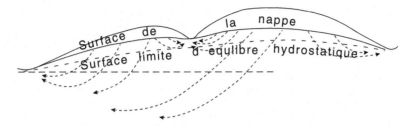

Fig. 1.2 'Allure complex des filets liquids dans une nappe libre': Complex pattern of fluid flow in an unconfined aquifer (Fourmarier, 1939, Figure 43, p. 87: after D'Andrimont, 1906).

conviction by the author concerning its existence or importance in nature. Both of these diagrams are perceptive generalizations of possible patterns of gravity-flow in complex drainage basins. However, they are only the mental images of individuals without the accompanying mathematical descriptions that would enable others to reproduce and further develop them.

The door to the transition from the age of speculative and qualitative conceptualizations of regional groundwater flow to rigorous mathematical analyses was opened by Hubbert's (1940) paper, *'The Theory of Ground-Water Motion'*. In this classic treatise, Hubbert derived the concept of fluid potential, Φ, from first principles. He also showed that, for subsurface liquids in general, Φ comprises two terms: one related to pore pressure, p, the other to topographic elevation, z, (because of low velocities the inertia-dependent kinetic energy associated with the fluid's motion can be neglected) and that the impelling force acting upon a unit mass of fluid is the negative first derivative of the fluid potential. Consequently, the force field can be calculated, or modelled, for any given flow domain along the boundaries for which Φ or its first derivatives can be stated. In turn, the flow field can be determined by combining the force field with the rock's hydraulic properties (porosity, permeability, storativity). Basinal-scale groundwater flow patterns could thus now be produced mathematically as solutions to formal boundary value problems. However, another twenty years went by before this gift to hydrogeology was exploited.

In the course of my regular duties as a hydrogeologist in Central Alberta, Canada, I noticed a discrepancy between what I expected on the basis of Hubbert's (1940) Figure 45, on the one hand (Fig. 1.3), and what I saw in the field, on the other.

Hubbert's figure showed all infiltrating water resurfacing in the thalweg of the valley, as though the watercourse were a drainage ditch. In reality, the beds of the numerous creeks in my area were dry in many places and whatever water they had was frozen to the bottom in the winter. Based on 'Figure 45', and considering the steep topographic slopes, shallow water tables (<3 m deep) and permeable rock, together providing sufficient supplies of water to the area's farms and towns, I

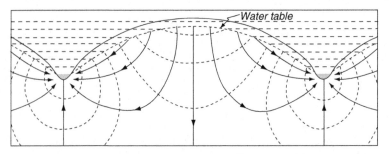

Fig. 1.3 'Approximate flow pattern in uniformly permeable material between the sources distributed over the air–water interface and the valley sinks' (Hubbert, 1940, Fig. 45, p. 930).

would have expected healthy runoffs in the creeks. A possible solution to the riddle occurred to me one day when I realized that the convergence of the flow lines in the figure was *a postulate, not a result;* Hubbert *made* the flow lines converge on the thalweg! I decided to find out where the water wants to go by itself, and solved the Laplace equation for a drainage basin of simple geometry (Fig. 1.4, App. A; Tóth, 1962a, Fig. 3, p. 4380).

The results were revealing: they showed that instead of the 'sinks' being 'limited to the bottoms of valleys containing streams' (Hubbert, 1940, p. 928), 'groundwater discharge is not concentrated in the valley bottom' (Tóth, 1962a, p. 4386). Thus the entire lower half of the basin was revealed to be a 'discharge area'. This simple discovery has triggered a number of follow-up studies in rapid succession.

During the preparation of the above, my first, paper (Tóth, 1962a), I already knew that assuming a linearly sloping valley flank was an oversimplification. I solved the Laplace equation again, now for a drainage basin with a sinusoidal surface superimposed on a linear regional slope (Fig. 1.4c; App. B; Tóth, 1962b, 1963 Fig. 3, p. 4807, reprinted in 1983). The analysis resulted in the groundwater flow-pattern for composite basins with homogeneous and isotropic rock framework. It was aptly called the ' hierarchically nested flow systems' by Engelen (Engelen and Jones, 1986, p. 9).

By fortunate coincidence, at the time when numerical methods just started to gain popularity R. Allan Freeze was looking for a Ph.D. thesis topic at the University of California, Berkeley. Advised by P. Witherspoon, he intended to show the value of the method to groundwater-related problems. Freeze took off from my solution to the composite-basin problem and produced a trail-blazing series of three papers from his thesis showing that quantitative flow-nets can be calculated by numerical, as opposed to analytical, methods for gravity-driven groundwater flow in drainage basins of arbitrary topography and heterogeneous and anisotropic rock framework (Freeze and Witherspoon, 1966, 1967, 1968).

(a)

(b)

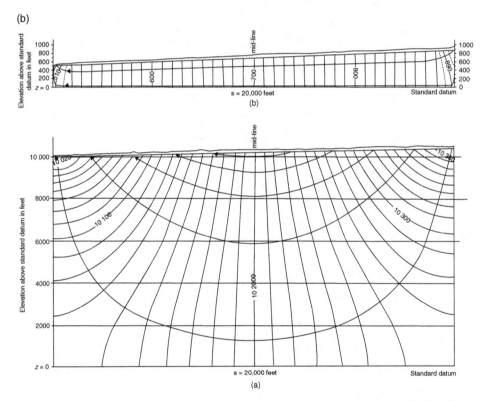

Fig. 1.4 Groundwater flow in a simple drainage basin: (a) the concept-inspiring topography, Central Alberta, Canada (photo by J. Tóth); (b) two-dimensional theoretical fluid-potential distributions and flow patterns for different depths to the horizontal impermeable boundary in a drainage basin with linearly sloping water table (Tóth, 1962a, Fig. 3, p. 4380). (c) Hierarchically nestled gravity-flow systems of groundwater in drainage basin with complex topography (Tóth, 1963, Fig. 3, p. 4807).

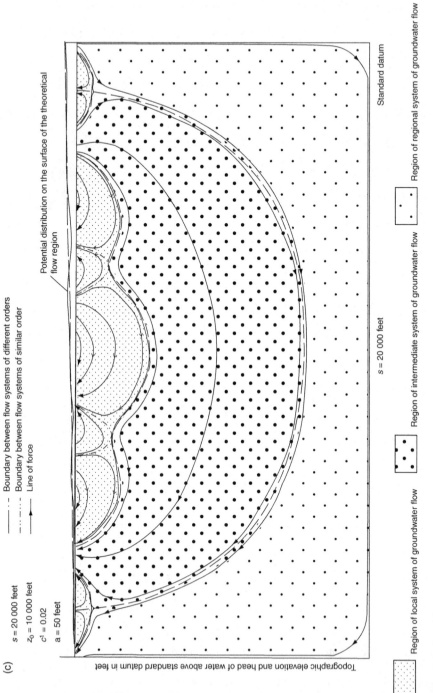

(c)

s = 20 000 feet

z_0 = 10 000 feet

c^1 = 0.02

a = 50 feet

——— · — Boundary between flow systems of different orders

——— · · — Boundary between flow systems of similar order

———▶ Line of force

Potential distribution on the surface of the theoretical flow region

Topographic elevation and head of water above standard datum in feet

Standard datum

s = 20 000 feet

Region of local system of groundwater flow

Region of intermediate system of groundwater flow

Region of regional system of groundwater flow

Fig. 1.4 (cont.)

Collectively, the above papers, published between 1962 and 1968, have fundamentally affected the direction, the scope and the rate of the subsequent evolution
of hydrogeology. In the search for and the development of groundwater resources,
basin-scale considerations have been added to aquifer- and well-hydraulics. The
scope of the field was broadened by dedicated field studies, recognizing moving groundwater as a geologic agent of diverse consequences and introducing
the flow-system concept into a wide variety of other disciplines such as soil science, petroleum exploration, economic geology, geothermics, hydrogeochemistry,
soil mechanics, sedimentology, diagenesis and ecology. Hydrogeology has thus
been turned into both a basic and a specialty discipline of the earth and hydrological sciences. In addition, the broadened scope of hydrogeologically-related
activities attracted talented new researchers and practitioners who accelerated the
rates of theoretical progress and methodological innovations. Some of the many
studies inspired by the flow-system concept and prompting further developments
include: Meyboom *et al.* (1966), Tóth (1966a, 1978, 1980), Fritz (1968), Mifflin
(1968), Williams (1968, 1970), Freeze (1969), Freeze and Harlan (1969), Kiraly
(1970), Deere and Patton (1971), Domenico and Palciauskas (1973), Schwartz and
Domenico (1973), Galloway (1978), Winter (1978), Garven and Freeze (1984),
Garven (1989) and so on.

Regional, or basinal, groundwater flow can be studied, characterized, and
evaluated by three different methods: (i) mathematical modelling; (ii) field measurements of fluid-dynamic parameters; and (iii) mapping of flow-generated natural
field-phenomena.

(i) Mathematical models produce spatially and, in the case of Equation (1.1) also temporally, distributed patterns of flow-related fluid-dynamic parameters. The patterns
can be obtained as solutions to the *Diffusion Equation* for transient, or non-steady-
state, flow:

$$\frac{\partial^2 h}{\partial x^2} + \frac{\partial^2 h}{\partial y^2} + \frac{\partial^2 h}{\partial z^2} = \nabla^2 h = \text{div grad } h = \frac{S_0}{K} \frac{\partial h}{\partial t} \qquad (1.1)$$

or as solutions to its particular case, the *Laplace Equation,* for steady-state flow (1.2):

$$\frac{\partial^2 h}{\partial x^2} + \frac{\partial^2 h}{\partial y^2} + \frac{\partial^2 h}{\partial z^2} = \nabla^2 h = \text{div grad } h = 0. \qquad (1.2)$$

In these equations h is hydraulic head, t is time, K is hydraulic conductivity and S_0 is
specific storage. The equations can be solved analytically, numerically or by analogue
modelling, using appropriate initial and boundary conditions.

(ii) Fluid-dynamic parameters [hydraulic heads, h; vertical gradients of pore pressure
or pressure vs. depth profiles, $p(d)$; and dynamic pressure-increments, Δp] can be

obtained or derived from field measurements of pore pressures and/or groundwater levels, and interpreted as patterns of fluid potential and flow.

(iii) Many different processes and phenomena are in cause-and-effect relation to gravity-driven groundwater in the realm of hydrology, ground- and/or surface-water chemistry, plants and plant ecology, mineralogy, pedology, soil- and rock-mechanics, subsurface transport of heat and mass, and so on. The manifestations of these natural conditions can thus be interpreted in terms of direction and intensity of flow.

1.2 Portrayal of groundwater flow-systems

1.2.1 Darcy's experiment and Law

The quantitative relation between the strength and sense of the fluid-driving force and the rate and direction of the flow that it induces through permeable materials was first stated, based on laboratory experiments, by the French engineer Henry Darcy (1856). This empirical relation is the fluid-flow equivalent of Ohm's Law for electrical current or Fourier's Law for heat flow. Because it encapsulates the main aspects of the physics of flow, is stated in terms of basic fluid-dynamic parameters and requires the use of essential terminology, Darcy's Law is the natural introduction to any discussion of subsurface fluid flow.

In Darcy's experiments, water was passed through a vertical iron pipe filled with sand and equipped with manometers along its side (Fig. 1.5).

The size of the sand grains and the rate of water flow were varied in the different experiments. The changes resulted in variation of the water levels in the manometer

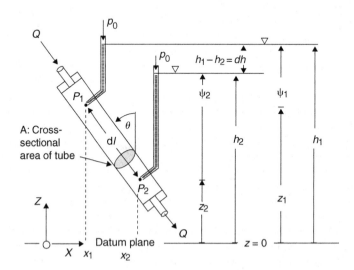

Fig. 1.5 Apparatus (with tilted pipe) to demonstrate Darcy's Law and the meaning of hydraulic (potentiometric) head.

tubes as measured vertically above a horizontal datum plane. The summary conclusion that Darcy made from his observations, which is called 'Darcy's Law' in
his honour, can be expressed mathematically in various forms, for instance:

$$Q = -KA\frac{dh}{dl},$$ (1.3a)

$$q = -K\frac{dh}{dl}.$$ (1.3b)

The terms in Equations (1.3a) and (1.3b) are defined and named below (dimensions
are shown in brackets) and illustrated in Figure 1.5.

$A[L^2]$: *cross-sectional area* of flow field normal to the direction of flow; $Q[L^3/T]$
total volume of fluid passing through A during a unit length of time t, or *volume
discharge;* $q = Q/A[L^3/T]/[L^2] = [L/T]$: volume of fluid passing through a unit
cross-sectional area of A during a unit length of time, *specific volume discharge,
flux, Darcy velocity* or *flow strength;* $h[L]$: height to which the fluid rises above
the datum plane of elevation $z = 0$, from an observation point P in the flow field,
i.e. from a manometer's intake, *hydraulic head;* $dl[L]$: distance measured along
the flow path between points in which hydraulic head is determined, *flow length;*
dh: difference in hydraulic heads determined in different points separated by the
distance dl along the flow path; dh/dl $[L/]/[L] = [L_0]$, change in hydraulic
head over a unit length of flow path, *hydraulic gradient,* taken positive in the
direction of increasing hydraulic head; $K[L^3/TL^2][L/L] = [L/T]$: a constant of
proportionality found by Darcy to depend on the grain size of the sand, it represents
the volume discharge during a unit length of time through a unit cross-sectional
area normal to flow, under a unit change in hydraulic head over a unit length of
flow path, *hydraulic conductivity;* the negative sign in the equation is used by
convention, in order to obtain a positive value for the volume discharge in the
direction of decreasing hydraulic head, i.e. in the direction opposite to the hydraulic
gradient.

Darcy conducted his experiments with descending flow in vertically positioned
columns. Nevertheless, his conclusions as expressed by Equations (1.3a) and (1.3b)
are valid for any flow direction relative to vertical, including horizontal and uphill
flow. Some of the more important observations that can be made immediately from
these equations are:

(i) flow is always in the direction of decreasing hydraulic head;
(ii) the volume discharge Q is directly proportional to the flow-field's cross-sectional area
A, the hydraulic conductivity K and the hydraulic gradient dh/dl;
(iii) the flux q is directly and linearly proportional to both the hydraulic conductivity K and
the hydraulic gradient dh/dl;

(iv) although the flux q has dimensions of velocity (hence the term: Darcy velocity) it represents a volumetric discharge rate;

(v) similarly, the hydraulic conductivity K has dimensions of velocity but in reality it is the specific flux that occurs under a unit hydraulic gradient;

(vi) since K is the flux normalized to a unit hydraulic gradient $dh/dl = 1$, its magnitude expresses the ease with which the fluid passes through the permeable medium.

1.2.2 Fluid-dynamic parameters

The fluid-dynamic parameters are variables with magnitudes and distribution in space and time that define and characterize the fields of flow and driving force. Consequently, they can be used graphically and/or numerically to map and analyse the intensity and direction of groundwater flow.

1.2.2.1 Fluid potential, Φ, and hydraulic head, h

The *fluid potential, Φ*, and the *hydraulic head, h*, are fluid-dynamic parameters whose physical meaning can be understood from Darcy's Law. One of the fundamental conclusions drawn from Darcy's experiments is that flow is not controlled either by elevation or by pressure exclusively. Figure 1.6 shows fluid pressures p to be proportional to the height of the water column ψ above measurement points P of elevation z in manometers that are open at the top. Accordingly, flow takes place from higher to lower elevation (from z_1 to z_2) but from lower to higher pressure

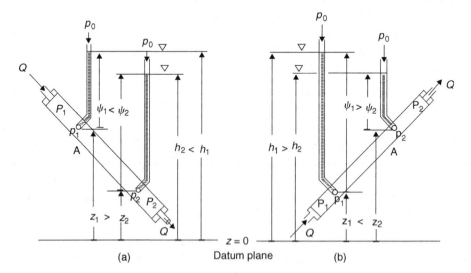

(a)   Datum plane   (b)

Fig. 1.6 Illustration of hydraulic-head change as the unique control on the direction of water flow: (a) flow from low to high pressure; (b) flow from low to high elevation.

(from p_1 to p_2) in Figure 1.6(a). On the other hand, flow occurs in the direction of decreasing pressures (from p_1 to p_2) but uphill (from z_1 to z_2) in Figure 1.6(b). In both cases, however, the flow is oriented towards declining hydraulic heads ($h_1 > h_2$).

Recognizing that in physical systems movements and processes always occur in the direction of decreasing energy, it is reasonable to postulate that the hydraulic head must somehow be related to the fluid's mechanical energy and may thus be a measure of it. In order to test this hypothesis, as well as to elucidate the physical nature of h, Hubbert's (1940) thought experiment is reviewed briefly below.

An elementary fluid mass m is considered in point P_1 of a subsurface environment with its physical state being characterized by the following initial conditions: elevation $z = z_1$; volume $V = V_1$; density $\rho = \rho_1$; velocity $v = v_1$; and pressure $p = p_1$. A change in any of these conditions requires a certain amount of work to be done on, or by, the fluid element m. In turn, the work thus performed will result in a gain or loss, respectively, of an equal amount of mechanical energy possessed by the fluid mass m, energy being defined (Hubbert, 1940, p.798) as 'the amount of work required to effect any given transformation from some arbitrary initial state to a specified final state'. If the fluid mass m is allowed to move from point P_1 to point P_2, it will be characterized in P_2 by the new set of physical conditions: z_2, V_2, ρ_2, v_2 and p_2. The work required to effect the changes of state from the first set of conditions to the second set can be resolved into three components.

First, work is required to lift the fluid mass m against gravity from z_1 to z_2. The potential energy of the fluid mass will thus be increased by the work increment:

$$w_1 = mg(z_1 - z_2) = mg \, dz. \tag{1.4}$$

Second, work is required to accelerate the fluid mass m from v_1 to v_2. This change in velocity increases the fluid's kinetic energy by the work increment:

$$w_2 = \frac{m}{2}(v_1^2 - v_2^2) = \frac{m\,dv^2}{2}. \tag{1.5}$$

Third, elastic energy is expended by relaxation of the internal stresses of the fluid mass due to a decrease in pressure from p_1 to p_2. This relaxation, which in the case of compressible fluids is accompanied by an increase in volume and a decrease in density, is equivalent to the work increment (Hubbert, 1940):

$$w_3 = m \int_{p_1}^{p_2} \frac{V_1 - V_2}{m} \, dp = \int_{p_1}^{p_2} \frac{dp}{\rho}. \tag{1.6}$$

Commonly, energy conditions are referenced to standard states, or datums. The conventional standard states used in subsurface hydraulics for the above energy types are for potential energy: sea level, $z = 0$; for kinetic energy: rest, $v = 0$; and for elastic energy: atmospheric pressure: $p = p_0$. Using these datums and notations to characterize the initial conditions of a fluid element m, the total work performed on it, i.e. the change in its mechanical energy content, is $W = w_1 + w_2 + w_3$. If the *fluid potential* is defined as the work performed on, or by, a *unit mass of fluid*, then the energy content or fluid potential of a unit mass that moved from P_1 to P_2 is:

$$\Phi = \frac{W}{m} = gz + \frac{v^2}{2} + \int_{p_0}^{p} \frac{\mathrm{d}p}{\rho}. \tag{1.7}$$

Based on the approximations that flow velocities in the subsurface are negligibly low ($v \approx 0$), and that most subsurface fluids may be considered only slightly compressible so that the density is not a function of pressure p ($\rho \approx$ constant) a simplified expression for the fluid potential is obtained:

$$\Phi = gz + \frac{p - p_0}{\rho}. \tag{1.8}$$

The pressure in a point, P, at the base of the manometers is

$$p = \rho g \psi + p_0, \tag{1.9}$$

where ψ is the height of the fluid column above P, and p_0 is atmospheric pressure on the fluid's surface in the manometer. Considering that $h = z + \psi$, Equation (1.9) can be written as $p = \rho g(h - z) + p_0$. Substituting into Equation (1.8) yields

$$\Phi = gz + \frac{[\rho g(h - z) + p_0] - p_0}{\rho} \tag{1.10}$$

or simply

$$\Phi = gh. \tag{1.11}$$

The simple relation of Equation (1.11) proves the validity of the initial hypothesis. The hydraulic head h is directly and linearly proportional to the fluid potential Φ through the gravitational constant g; h is thus a measure of the mechanical energy content of the fluid. The use of h for the quantitative characterization of flow rates and directions is, therefore, possible and physically justified.

It is common in subsurface hydrology to consider atmospheric pressure as a datum (mainly because pressure gauges usually read values with reference to the atmosphere) and thus to refer to the measured pressures as gauge pressure. In such cases, the atmospheric pressure $p_0 = 0$, and by combining Equations (1.8), (1.9) and (1.11):

$$\Phi = gh = gz + \frac{p}{\rho} \qquad (1.12)$$

or

$$h = z + \frac{p}{\rho g} \qquad (1.13)$$

and

$$h = z + \psi. \qquad (1.14)$$

Equation (1.14) was anticipated by Darcy's experiment as illustrated, e.g. in Figures 1.5 and 1.6. It is, however, filled now with physical content: h is a direct measure of the fluid's potential energy in a given point of the flow region (Eq. 1.11). According to the relation, the *hydraulic head,* or *potentiometric elevation,* h consists of two components: (i) the *elevation head,* z, or the height of the point of observation, P, above the datum plane and (ii) the *pressure head,* ψ, which is the height of the fluid column above point P. (Occasionally the term 'piezometric head' is also used for h. Verbatim it means 'pressure measuring'. It is, therefore, a misnomer because h is proportional to the total *energy,* and *pressure* is only one of its two components.) It should also be noted that the fluid potential was defined for fluids of constant density, ρ. The hydraulic head h can, therefore, be used as a measure of fluid-potential energy in homogeneous fluids only.

Occasionally some conceptual difficulties arise from the fact that the hydraulic head h is measured with respect to elevation $z = 0$, yet it represents the fluid potential in a point which, in general, is not in the datum plane. An irrelevance may thus be perceived between the place of measurement (e.g. pressure gauge in an oil well, from which the hydraulic head is calculated) and the place of reference, i.e. the datum plane of $z = 0$ (Figure 1.7).

This question becomes important in the evaluation of fluid-flow patterns from measured or calculated distributions of hydraulic heads. Figure 1.7 illustrates the point. Significantly different pressures are indicated by the differing pressure heads $\psi_1 \ll \psi_2$ in points P_1 and P_2. Nevertheless, flow cannot occur between the two points because their respective energy levels are equal: $h_1 = h_2$. In point P_1 the energy due to pressure and represented by ψ_1 is deficient relative to $z = 0$, as indicated by h_1. The deficit is compensated for, however, by the positive elevation head z_1, i.e. by the energy of position. On the other hand, the high energy of pressure indicated by ψ_2 in point P_2 cannot be converted into a driving force relative to P_1 because of the negative z_2 position of the point of measurement. A

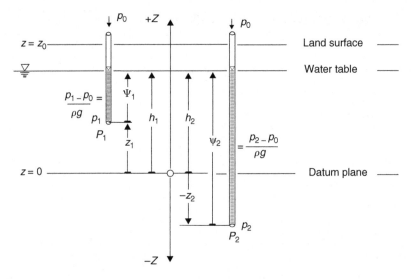

Fig. 1.7 Interpretation of the hydraulic (potentiometric) head in the $+z$ and $-z$ half spaces.

clear understanding and appreciation of the above relations are indispensable in the construction and interpretation of the potentiometric distributions (surfaces and cross-sections), which are basic working tools of the hydrogeologist.

1.2.2.2 Pore pressure vertical pressure-gradient and dynamic pressure-increment

Pore pressure in a saturated subsurface water domain open to the atmosphere was described earlier as the pressure due to the combined weight of the fluid column and atmosphere at the base of a manometer. Defined generally, also including closed systems, *pore pressure, p,* is the mechanical force acting on the fluid per unit surface area in the rock's pores at a given point in the subsurface. Other terms also used include *formation pressure, formation-fluid pressure* and *neutral stress.*

   Pore pressures may be generated by various mechanisms, and their vertical and horizontal variations may be affected by several factors. The concepts of pore pressure and the dynamic parameters derived from it (namely, the vertical pressure-gradient and dynamic pressure-increment) are introduced in this section based on vertically one-dimensional analysis and a homogeneous flow medium. Regional patterns of pore pressures and their controlling influences in different geologic environments will be discussed in later sections.

(i) Pore pressure and hydrostatic pressure-gradient. In subsurface flow problems formation pressures and their gradients are commonly presented and analysed

in Cartesian coordinate systems, with vertical Z and horizontal X and Y axes. Two different reference levels can be used for the origin of the coordinate system (Fig. 1.8a). In one, the more commonly used $p(z)$, system the origin of the XY *horizontal datum* plane is at sea level, $z = 0$, and vertical positions thus correspond to topographic altitudes. The Z-axis is taken positive upward. The pressures p, or pressure heads ψ, are plotted on a horizontal axis. In the other, the $p(d)$, system the vertical axis represents depth d below *land surface* and its origin is at $d = 0$. The depth axis is taken positive downward (Fig. 1.8a). Here too, the horizontal axis represents pressures or pressure heads. The *elevation of the water table* above the $p(z)$ system's datum plane is denoted z_0, and its depth below the land surface is d_0.

A subsurface fluid environment is termed *unconfined* if the upper boundary of the fluid body contained in it is a free surface at which the pressure is atmospheric, p_0. The surface in the groundwater body at which the pressure is p_0 is called the *water table*. The space between the land surface and the water table is the *unsaturated zone, vadose zone* or *zone of aeration*.

In practice, pressures are expressed either as absolute, or as atmospheric or gauge pressure. *Absolute pressure* p_a is referenced to vacuum, i.e. to a pressure of absolute zero. *Gauge pressure* p, on the other hand, is measured relative to the pressure of the atmosphere, p_0, i.e. to one atm., or approximately 100 kPa. The difference between the two pressures is, therefore, one atm., and they are related to one another as: $p = p_a - p_0$. (Fig. 1.8b). In this book, pore pressures are expressed as gauge pressures and denoted as p.

The vertical gradient of the pore pressure, or *vertical pressure-gradient, γ*, is the change in pore pressure per unit vertical length in a flow domain. It controls the direction and intensity of the vertical flow component and is, therefore, a useful parameter in the analysis and interpretation of various aspects of the flow field. The concept can be conveniently illustrated by considering the pressures in two subsurface points at different elevations along a vertical.

The pressures p_1 and p_2 in the points P_1 and P_2 are (Fig. 1.8a):

$$p_1 = \rho g \psi_1 + p_0 = \gamma \psi_1 + p_0, \tag{1.15a}$$

$$p_2 = \rho g \psi_2 + p_0 = \gamma \psi_2 + p_0, \tag{1.15b}$$

where ρ is the fluid's density, g is the gravitational constant, and $\gamma = \rho g$ is the fluid's specific weight, other terms having been previously defined. Taking the difference between the two pressures and dividing by the vertical distance between P_1 and P_2, $dd = -dz$ (Fig.1.8b), yields: $(p_2 - p_1) / dd = dp / dd = \rho g (\psi_2 - \psi_1) dd = \gamma \, d\psi / dd$.

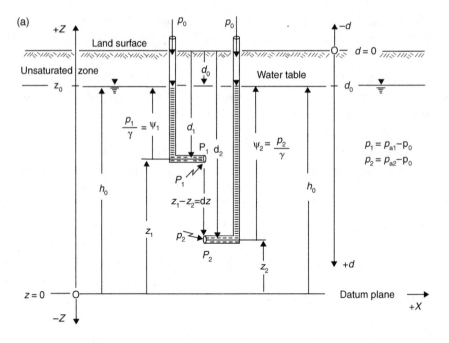

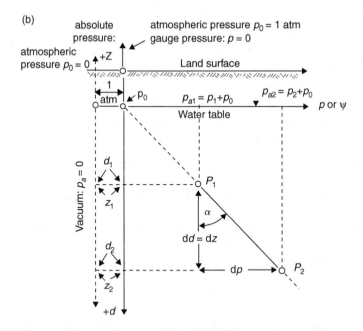

Fig. 1.8 Interpretation and presentation of static pore-pressures in unconfined environments: (a) notations; (b) pressure versus depth, $p(d)$, profile.

In a hydrostatic environment, however, $d\psi = dd = -dz$ (Fig. 1.8b). The vertical gradient of the hydrostatic pore pressure in a homogeneous unconfined fluid is therefore:

$$\frac{dp}{dd} = -\frac{dp}{dz} = \rho g = \gamma. \tag{1.16a}$$

Thus the rate of vertical pressure change in a static fluid numerically equals the specific weight of that fluid. However, if the fluid is in motion and the flow has a vertical component, this equality is not valid, as will be shown later. Sometimes, therefore, it may be necessary to state explicitly that the specific weight γ in question reflects the fluid's density, ρ, i.e. that it has not been affected by vertical flow. In such cases it is advisable to indicate the static nature of γ explicitly by a suffix 'st': γ_{st}.

The correctness of the relation in Equation (1.16a) can be verified by simple dimensional analysis. Since pressure is force per surface area, the dimensions of the pressure gradient are: $\frac{[dp]}{[dd]} = \frac{[MLT^{-2}][L^{-2}]}{[L]} = \frac{[M]}{[T^2L^2]}$. In anticipation of some of the practical uses of the above relations in deep-basin hydrology, we may already state that a vertical pressure gradient that is numerically equal to the fluid's specific weight is suggestive, but not a proof (!), of a static and unconfined fluid system.

In practice, vertical pressure-gradients are diagrammed by plotting pore pressures against elevation or depth of the observation points in two-dimensional co-ordinate systems (Fig. 1.8b). Depending on various considerations to be discussed later, the plotted points are connected individually or averaged. The procedure yields the *vertical pressure-gradient profile* or *pore-pressure profile*. If the pressures are plotted against elevation, the resulting curve is a *pressure-elevation profile* or a $p(z)$ profile; if depth is the independent variable, the curve is called a *pressure-depth profile* or a $p(d)$ *curve* (Fig. 1.8b).

From Figure 1.8b it can also be seen that the *slope* of a $p(z)$- or $p(d)$- *curve* is a direct expression of the vertical pressure-gradient:

$$\tan \alpha = \frac{dp}{dd} = -\frac{dp}{dz} = \gamma. \tag{1.16b}$$

The slope can thus be interpreted conveniently, even visually, in terms of vertical gradients of measured pore pressures. The procedure is used commonly for comparing and analysing subsurface flow in different areas or at different depth ranges. Table 1.1 and Figure 1.9 present densities and vertical hydrostatic pressure gradients numerically and graphically for typical formation fluids. For comparison, the vertical pressure gradient in a rock column of average density, i.e. the *lithostatic pressure gradient*, is also included.

Table 1.1. *Vertical pressure-gradients for static subsurface fluids of typical densities*

Curve number on Fig. 1.9	Substance	Typical mineralization (TDS) (mg/l)	Characteristic density ρ (kg/m³)	Vertical static gradient $\tan \alpha = \gamma_{st}$	
				kPa/m	psi/ft
1	Natural gas	–	150	1.47	0.21
2	Crude oil	–	850	8.34	1.21
3	Pure water	0	1000	9.8067	1.42
4	Fresh water	10 000	1010	9.84	1.43
5	Saline (brackish) water	150 000	1105	10.86	1.58
6	Brine	320 000	1260	12.37	1.79
7	Indurated rocks	–	2700	26.47	3.84

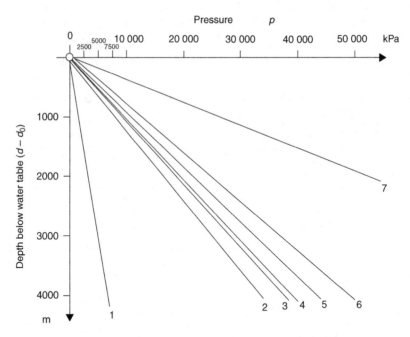

Fig. 1.9 Hydrostatic pressure–depth profiles for subsurface fluids of various densities; pure water and lithostatic loading are added for reference.

(ii) Dynamic pressure increment and dynamic pressure gradient Darcy's experiment showed (Fig. 1.5) that for flow to occur between two points, the hydraulic heads, h, must be different in those points. But it also showed that, other things

remaining equal, a change in flow rate results in changes in hydraulic heads h in each point of the flow field. Consequently, in a single point of fixed elevation z a change in hydraulic head can occur only through a change in the pressure p, i.e. pressure head ψ, and it indicates a change in flow relative to the previous condition. Thus, in a given flow domain, the pressures and hydraulic heads (their values and spatial patterns) of the static state are different from those of the dynamic state. Because the difference is a function of the direction and intensity of the flow, it reflects on certain features of the flow field and can thus be useful in its analysis and characterization. The parameter expressing the difference in pore pressures between the states of flow and rest in one point of observation is the dynamic pressure increment, Δp.

The *dynamic pressure increment, Δp,* is defined, therefore, as the difference between the real or actual dynamic pressure, $p_{real} = p_{dyn}$, and the nominal or hypothetically static-state pressure, $p_{nom} = p_{st}$, in a single point P. Thus, $\Delta p = p_{real} - p_{nom} = p_{dyn} - p_{st}$. The parameter thus expresses the pressure difference that is available to generate vertical flow towards or away from the water table at the point of consideration.

Theoretically, the definition is valid and the use of this parameter is possible in three-dimensional space too. In practice, however, it has been applied to date in the vertical sense only, mainly due to the relative ease in comparing vertically spaced pressure measurements obtained in single bore holes or plotted on $p(z)$ profiles.

The concept and potential use of the dynamic pressure increment may be demonstrated by changes in the fluid-dynamic parameters due to the transition from a hydrostatic to a hydrodynamic state in one point of the fluid regime. Figure 1.10 shows a homogeneous unconfined subsurface regime of a fluid of constant density, ρ. Formation pressures or water levels are observed at point P at elevation z. A thought experiment is conducted on this regime in two parts. In state I, illustrated on the left-hand side of Figure 1.10, the fluid body is static and its energy conditions are characterized by the static hydraulic head:

$$h_{st} = z + \psi_{st} = z + \frac{p_{st}}{\rho g} = z + \frac{p_{st}}{\gamma_{st}}, \tag{1.17a}$$

where the subscript st indicates values at hydrostatic conditions.

In state II, illustrated on the right-hand side of Figure 1.10, the fluid is allowed to flow vertically downward by opening the well bottom at point P. In the meantime, the water table is kept constant in its original position by, for instance, precipitation. However, maintaining a vertical flow component requires a hydraulic gradient,

Introduction

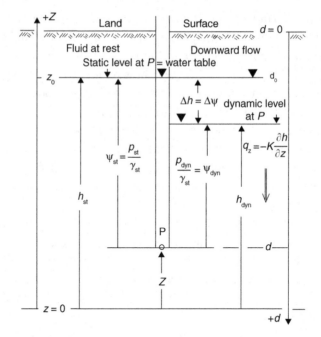

Fig. 1.10 Illustration of hydrodynamic parameters for the definition of the dynamic pressure increment, Δp.

i.e. change in the hydraulic head in point P, according to Darcy's Law, namely: $q_z = -K\partial h/\partial z$. (The partial differential operator ∂ means simply that only the vertical component of the possible total hydraulic gradient is considered.) In the new, dynamic, situation the energy conditions are characterized by the dynamic hydraulic head:

$$h_{dyn} = z + \psi_{dyn} = z + \frac{p_{dyn}}{\rho g} = z + \frac{p_{dyn}}{\gamma_{st}}. \qquad (1.17b)$$

The subscript dyn indicates that the parameter refers to a dynamic, i.e. flowing, condition.

The *dynamic head increment,* i.e. the difference in hydraulic heads representing the flowing and the static conditions, is obtained from Equations (1.17a) and (1.17b):

$$\Delta h = h_{dyn} - h_{st} = (z - z) + \frac{p_{dyn} - p_{st}}{\gamma_{st}} \qquad (1.18a)$$

from which the dynamic pressure increment is

$$\Delta p = \gamma_{st} \cdot \Delta h = p_{dyn} - p_{st}. \qquad (1.18b)$$

The static-pressure term p_{st}, in Equation (1.18b) can be calculated if the fluid's density ρ, ($\gamma_{st} = \rho g$), the depth d to point P, and the depth d_0 to the water table are known:

$$p_{st} = \gamma_{st} \psi_{st} = \gamma_{st}(d - d_0). \tag{1.19}$$

In other words, the static pressure p_{st} can be calculated regardless of whether the actual condition is static or dynamic. Such a calculated value of a hypothetically static pressure is a nominal hydrostatic pressure: $p_{st} = p_{nom}$.

The dynamic pressure p_{dyn}, on the other hand, can be actually determined by measuring the pressure in point P or the water level in an open well. The dynamic pressure increment is, therefore, a physical quantity that may be numerically evaluated, namely as

$$\Delta p = p_{dyn} - \gamma_{st}(d - d_0). \tag{1.20a}$$

In most real-life situations the depth, d_0, to the water table is small as compared with the depth, d, of the point of observation P and it may thus be neglected to yield

$$\Delta p = p_{dyn} - \gamma_{st} \cdot d = p_{real} - p_{nom}. \tag{1.20b}$$

The magnitude of Δp is proportional to the strength of the vertical flow component and it is dependent also on the rock's hydraulic conductivity K. Its sense, on the other hand, is indicative of the direction of the driving force (thus of the vertical component of the flow q_z in point P *with respect to* the water table) as follows:
if $p_{dyn} < p_{st}$ then Δp is *negative* ($\Delta p < 0$) and $q_z < 0$, i.e. flow is *downwards* with respect to the water table;
if $p_{dyn} > p_{st}$ then Δp is *positive* ($\Delta p > 0$) and $q_z > 0$, i.e. flow is *upwards* with respect to the water table.

The dynamic pressure-increment can be used in real-life situations to determine if vertical flow exists relative to the water table. It is done by comparing actually measured $p(d)$-profiles with $p(d)$ plots of pressures calculated for the fluids of known, or probable, densities of the region (Fig. 1.11). Figure 1.12 presents an alternative way to plot the information shown in Figure 1.11. In the latter case, the hydrostatic pressure curve, serving as reference, is plotted as a vertical line. The calculated dynamic pressure-increment values are then plotted on the abscissa. Positive or negative departures from the vertical indicate the vertical sense of fluid flow with respect to the water table for the regions in which the measurements were obtained.

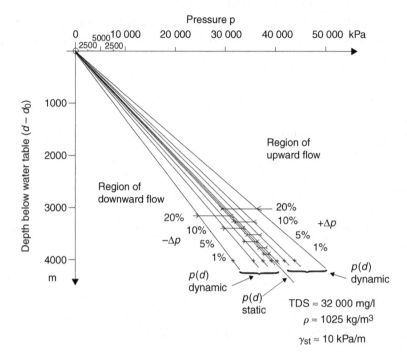

Fig. 1.11 Theoretical $p_{dyn}(d)$-profiles, and dynamic pressure increments, Δp, attributed to vertical downward flow $(-\Delta p)$ and upward flow $(+\Delta p)$ components of formation water of nearly sea-water salinity, referenced to the water table, in a homogeneous and unconfined rock framework. Differences in relative magnitude of Δp may be caused by differences in either the rock's hydraulic conductivity or in the maximum hydraulic head values, or both.

1.2.3 The Laplace and diffusion equations

Groundwater flow is a physical process. Its theoretical distribution in space and time for specified flow domains can thus be determined mathematically. The general equation for flow patterns and intensities can be derived by combining Darcy's Law with the equation of continuity. Darcy's Law is an equation of momentum balance, while the continuity equation expresses a balance of mass, i.e. the principle or mass conservation. Specific situations are analysed as particular solutions to the general equation in the form of 'boundary-value' problems.

The derivation of the general equation is illustrated on the Laplace equation. It describes saturated flow under steady state conditions and is one of the most frequently used particular solutions in problems of regional groundwater flow.

To this end, the rate of mass-flow, or mass flux, is considered through an elemental cube of control volume, normal to its three faces, $\Delta x \Delta y$, $\Delta x \Delta z$ and $\Delta y \Delta z$ (Fig. 1.13):

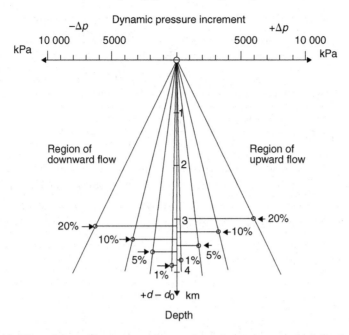

Fig. 1.12 Plot of dynamic pressure-increments, Δp, using nominal (hydrostatic) pressures as reference. Percentages indicate magnitude of the dynamic pressure increment as fraction of the nominal hydrostatic pressure at the depth of pressure measurement.

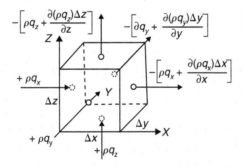

Fig. 1.13 Fluid mass flow through an elementary control volume, V, with dimensions of unit length; $V = x \times y \times z = 1 \times 1 \times 1$.

$+\rho q_x\, \Delta y \Delta z - (\rho q_x + \frac{\partial(\rho \cdot q_x)}{\partial x}\Delta x)\, \Delta y \Delta z,$ i.e. flow in the X direction,

$+\rho q_y\, \Delta x \Delta z - (\rho q_y + \frac{\partial(\rho \cdot q_y)}{\partial y}\Delta y)\, \Delta x \Delta z,$ i.e. flow in the Y direction and

$+\rho q_z\, \Delta x \Delta y - (\rho q_z + \frac{\partial(\rho \cdot q_z)}{\partial z}\Delta z)\, \Delta x \Delta y,$ i.e. flow in the Z direction,

where x, y, z are coordinates, ρ is fluid density, and q is specific volume discharge, or Darcy velocity.

In the case of steady-state flow the conservation of mass requires that the total flow into the volume element be equal to the total flow out of it, thus that the sum of changes in the three directions is zero. By this requirement the sum of terms of the form $+\rho q_x \, \Delta y \Delta z - \rho q_x \, \Delta y \Delta z = 0$, and the sum of the changes also must be zero. The latter statement yields the *equation of continuity:*

$$\Delta V \left[\frac{\partial(\rho \cdot q_x)}{\partial x} + \frac{\partial(\rho \cdot q_y)}{\partial y} + \frac{\partial(\rho \cdot q_z)}{\partial z} \right] = div(\rho \, \mathbf{q}) = 0. \tag{1.21}$$

Expanding the left-hand side according to the chain rule results in terms of the form $\rho \frac{\partial q_x}{\partial x}$ and $q_x \frac{\partial \rho}{\partial x}$. For slightly compressible fluids, terms of the first type are much larger than those of the second type. Consequently, ρ may be neglected both for the case when it varies in space, i.e. when $\rho(xyz) \neq$ constant, and also if $\rho(xyz) =$ constant.

Equation (1.21) thus simplifies to:

$$\frac{\partial q_x}{\partial x} + \frac{\partial q_y}{\partial y} + \frac{\partial q_z}{\partial z} = div \, \mathbf{q} = 0, \tag{1.22}$$

which is the equation of continuity for slightly compressible fluids (most liquids but not the gases).

This equation is combined now with Darcy's Law, in which the vector of specific volume discharge, $\mathbf{q}$, is resolved into its three components as, for instance, $q_x = -K_x \frac{\partial h}{\partial x}$.

The combination yields the equation for steady-state flow through an anisotropic permeable medium:

$$\frac{\partial}{\partial x} \left(K_x \frac{\partial h}{\partial x} \right) + \frac{\partial}{\partial y} \left(K_y \frac{\partial h}{\partial y} \right) + \frac{\partial}{\partial z} \left(K_z \frac{\partial h}{\partial z} \right) = 0. \tag{1.23}$$

For isotropic, $K_x = K_y = K_z$, and homogeneous, $K(x, y, z) =$ constant, media Equation (1.23) reduces to the *Laplace equation,* which thus represents the equation for steady-state flow through such rock types:

$$\frac{\partial^2 h}{\partial x^2} + \frac{\partial^2 h}{\partial y^2} + \frac{\partial^2 h}{\partial z^2} = \nabla^2 h = 0. \tag{1.2}$$

Solutions to this equation can be portrayed graphically by *flow nets,* which show the distribution of flow paths in the horizontal plane, $h(x, y)$ [*potentiometric surface map*], and vertical section, $h(x, z)$ [hydraulic cross-section]. By solving the Laplace equation analytically, values of h can be calculated for every point $P(x, y, z)$ of a continuous domain. Numerical solutions yield h values for discrete points. In both approaches, however, the geometry of the domain and h or grad$h$ along the boundaries must be specified to obtain solutions.

For non-steady state, or transient, flow the sum of inflow to and outflow from the control volume is not necessarily zero because the receptacle is able to store and/or release fluid due to the compressibility of rock and water. Flow is controlled in this case by the *diffusion equation:*

$$\frac{\partial^2 h}{\partial x^2} + \frac{\partial^2 h}{\partial y^2} + \frac{\partial^2 h}{\partial z^2} = \nabla^2 h = \frac{\partial h}{\partial t} \frac{S_0}{K},$$
(1.1)

where t is time, S_0 is specific storage and K is hydraulic conductivity.

2

The 'Unit Basin'

The Unit Basin is the elementary building concept in the theory of gravity-driven regional groundwater flow. It is envisaged as a two-dimensional vertical slab of unit thickness and of homogeneous and isotropic hydraulic conductivity of the Earth's crust (Fig. 2.1). On the top, it is bounded by an axially symmetrical topographic depression with water tables rising linearly from the central thalweg to the divides, by an effectively impermeable horizontal stratum at the base, and by two vertical planes forming the sides beneath the water divides. The sides are considered impervious to flow owing to the water table's symmetry relative to the divides. The *basic pattern* of regional groundwater flow is a steady-state flow-field, which develops in a saturated unit basin with gravity being the sole water-driving force. *General flow patterns* may be thought of as modified versions of the basic pattern. Commonly, general patterns can be resolved into cells or compartments in which flow characteristics are similar to those of the basic pattern.

2.1 The basic flow pattern

The Unit Basin as defined above is a conservative flow domain. In a conservative domain no sources or sinks of energy or flow occur. The distributions of hydraulic head and flow can thus be calculated by the Laplace equation (Eq. 1.2) using the appropriate boundary conditions (Fig. 2.1; App. A; Tóth, 1962a; Freeze and Witherspoon, 1967; Wang and Anderson, 1982; Zijl and Nawalany, 1993). Owing to the geometrical symmetry of the water table with respect to the Z-axis, the (y, z) vertical planes at $x = 0$ and $x = s$ may be considered as impermeable boundaries. Consequently, a flow pattern obtained for one flank of the basin is the mirror image of the flow distribution in the opposite flank. Due to the assumed impermeability of the basal and lateral boundaries, the components of the hydraulic gradients and flow normal to them are zero. The boundary conditions for the two sides and the

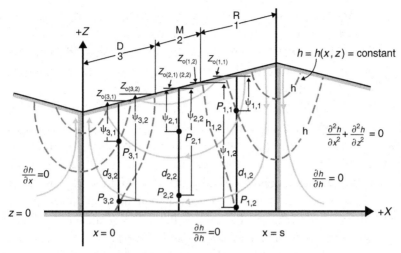

R, M, D : Area of Recharge, Midline, Discharge, respectively;
$P_{1,2}$: point of measurement in cased well; $d_{1,2}$: well depth; $\psi_{1,2}$: pressure head;
h : hydraulic head; $z_{o(1,2)}$: intersection of isopotential $h_{1,2}$ with water table.

Fig. 2.1 Patterns of hydraulic head, $h(x, z)$, and groundwater flow, $q(x, z)$, in the unit basin.

base of the basin are, therefore:

$$\left(\frac{\partial h}{\partial x}\right)_{x=0} = \left(\frac{\partial h}{\partial x}\right)_{x=s} = \left(\frac{\partial h}{\partial z}\right)_{z=0} = 0. \tag{2.1a}$$

The upper boundary of the flow field is the water table. Along the water table the hydraulic head h_{wt} is a linear function of the distance x from the centre of the basin:

$$h_{wt} = h_{wt}(x). \tag{2.1b}$$

The Laplace equation, combined with the four boundary conditions, yields the solution for the hydraulic head distribution in the unit basin as follows (Tóth, 1962a; App. A):

$$h = \left(z_0 + \frac{cs}{2}\right) - \frac{4gcs}{\pi^2} \sum_{m=0}^{\infty} \frac{\cos[(2m + 1)\pi x/s]\cosh[(2m + 1)\pi z/s]}{(2m + 1)^2\cosh[(2m + 1)\pi z_0/s]} \tag{2.2}$$

where $m = 1, 2, 3, \ldots, \infty$. With Equation (2.2) hydraulic head values h can be calculated for any point, $P_{x,z}$, in the unit basin characterized by depth z_0 at the thalweg, regional slope of the water table $c = \tan(\alpha)$, and width s of one flank. Equipotential lines can be obtained by connecting points of equal hydraulic heads. A complete flow net can be produced by drawing flow lines normal to the equipotential lines (in practice, flow nets are usually calculated by computer programs).

Figure 2.1 illustrates a unit basin with the basic patterns of groundwater flow and hydraulic head. Three functionally different flow-regimes can be distinguished in the basic patterns. They are the *hydraulic regimes of recharge* or *inflow, R; transfer or throughflow, M;* and *discharge* or *outflow,* D. These regimes are located in the areas, respectively, upslope, adjacent, and downslope, relative to the hydraulic midline. Relative to the water table, the direction of flow is descending, lateral, and ascending, respectively, in the three hydraulic regimes. The flow intensity, q, decreases with increasing depth and away from the midline area. In the lower corners of the flow field near-stagnant, or 'quasi-stagnant', conditions prevail. If the midline is interpreted in the strict sense, i.e. that it has no areal extent, the ratio of the area of recharge to the area of discharge is $R/D \approx 1$.

The physical meaning of the hydraulic-head distribution may be visualized by interpreting the potentiometric contours, $h(x, z) =$ constant, in terms of elevations to which water would rise in vertical bore-holes from points, $P_{x,z}$. These water levels can be determined by noting that the value of the hydraulic head in any point along a given contour line h_i is equal to the elevation of the water table at its intersection with the specified contour: $h_i = h_{wt} = z_0$. Consequently, water will rise from any point on a hydraulic head contour to the elevation at which that particular contour intersects the water table. Thus, in Figure 2.1, water from point $P_{1,1}$ will rise to elevation $z_{0(1,1)}$, while from point $P_{1,2}$ it rises to $z_{0(1,2)}$. Because elevation $z_{0(1,1)}$ is higher than $z_{0(1,2)}$, a downward sense of the vertical flow component is indicated. By similar considerations it can be determined that water levels in points $P_{2,1}$ and $P_{2,2}$ of the midline region are equal, namely, $z_{0(2,1)} = z_{0(2,2)}$, whereas water rises from the shallow point $P_{3,1}$ to the low elevation of $z_{0(3,1)}$ and from the deep $P_{3,2}$ to the higher $z_{0(3,2)}$.

Laterally, the hydraulic heads decrease from the recharge area towards the discharge area in any plane parallel either to the X-axis or to the water table. In plan view of a series of laterally repeated symmetrical basins, the divides appear as loci of hydraulic head maxima and diverging flow lines, whereas hydraulic heads are minimum at, and flow lines converging on, the thalweg.

In summary, the hydraulic head pattern of the unit basin is characterized by water levels that decline, are constant, and rise with depth in the recharge, hydraulic midline, and discharge areas, respectively. In map-view, the hydraulic heads are maximum at the basin's divide and minimum at the valley bottom. The hydraulic gradients decrease generally with increasing depth. Theoretically, they become zero in the two lower corners of the basin. Such regions of sluggish or zero formation-water flow can favour accumulation of natural mineral matter, anthropogenic waste or geothermal heat, and play a role in the entrapment of petroleum.

2.2 Basic patterns of fluid-dynamic parameters

In addition to the hydraulic head h, regional patterns of the pore pressure p, vertical pressure-gradient $\gamma = dp/dd = -dp/dz$ and the dynamic pressure increment Δp can also be used to characterize basinal flow conditions. The concepts, definitions and physical interpretation of these fluid-dynamic parameters were discussed in Chapter 1, based on one-dimensional flow. The spatial distribution patterns of the same parameters are interpreted in a two-dimensional basinal context below.

2.2.1 Pore pressure: p

In the unit basin the nominal static pressure-head equals the depth d_{xz} below the water table, $\psi_{st} = d_{xz}$, in any point P_{xz} (Figs. 2.1, 2.2a). On the other hand, the dynamic pressure head $\psi_{(xz)dyn}$, can be determined as the difference in elevations between the point of measurement z_{xz}, and the point of intersection of the water table with the hydraulic head contour passing through the measurement point (Figs. 2.1, 2.2a).

A systematic pattern of dynamic pore pressures emerges by comparing dynamic pressure heads with values of nominal static pressure heads in arbitrarily selected points of the three flow regimes as, for instance, (Figs. 2.1, 2.2a):

Site 1, recharge area: $\psi_{(12)dyn} < d_{12}$, or, in general: $\psi_{(xz)dyn} < d_{xz}$.

$$\text{Therefore:} \quad p_{(xz)dyn} = \gamma_{st}\psi_{(xz)dyn} < \gamma_{st}d_{xz} = p_{(xz)st}. \tag{2.3a}$$

Site 2, midline area: $\psi_{(22)dyn} = d_{22}$ or, in general, $\psi_{(xz)dyn} = d_{xz}$.

$$\text{Therefore:} \quad p_{(xz)dyn} = \gamma_{st}\psi_{(xz)dyn} = \gamma_{st}d_{xz} = p_{(xz)st}. \tag{2.3b}$$

Site 3, discharge area: $\psi_{(32)dyn} > d_{32}$ or, in general, $\psi_{(xz)dyn} > d_{xz}$.

$$\text{Therefore:} \quad p_{(xz)dyn} = \gamma_{st}\psi_{(xz)dyn} > \gamma_{st}d_{xz} = p_{(xz)st}. \tag{2.3c}$$

Accordingly, the wells of increasing depth in the areas of recharge, midline and discharge are terminated in parts of the basin where the fluid-potentials, respectively, decrease, remain constant and increase with depth (Fig. 2.2a).

Thus it appears that in those, but only those, areas where the flow has a vertical component the dynamic pore pressures are different from what the pore pressures would be at the same place and depths if there were no flow. In particular, the dynamic pore pressures p_{dyn} are less than, equal to, and greater than the static or nominal pore pressures $p_{st} = p_{nom}$ in the recharge, midline and discharge areas, respectively. The absolute values of the differences increase in both directions away from the midline and with increasing depth. They are zero at the midline through the basin's entire depth, and maximum in the two lower corners of the flow field.

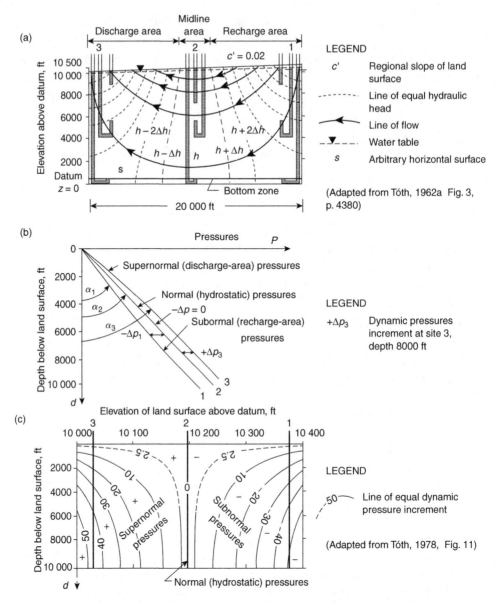

Fig. 2.2 Relation between fluid dynamic parameters in the unit basin: (a) hydraulic head, $h(x, z)$ and flow, $q(x, z)$; (b) pore pressure, $p(d)$; and (c) dynamic pressure increment, $\Delta p(z_0, d)$ (Tóth, 1980, Fig. 2, p. 124).

2.2.2 *Vertical pressure-gradient:* $dp/dd = -dp/dz = \gamma$: *pressure-vs.-depth or $p(d)$-profile*

The systematic distribution of pore pressures in the unit basin can be plausibly shown by means of a family of pressure versus depth curves, or $p(d)$, profiles.

The three curves 1, 2 and 3 in Figure 2.2b represent the vertical distributions of the dynamic pressures at sites 1, 2 and 3, respectively. Pressure plots between the numbered sites would be fanning between the numbered lines. The pressure increases with depth at each site, but at different rates. The average rate of pressure increase, $dp/dd = \gamma$, is minimum, and less than hydrostatic, beneath recharge areas: $\gamma_{1dyn} < \gamma_{st}$; it equals the hydrostatic rate of increase in the midline regions, $\gamma_{2dyn} = \gamma_{st}$; and it is maximum, and greater than the hydrostatic rate, in the areas of upward flow, $\gamma_{3dyn} > \gamma_{st}$.

Recalling that $\tan \alpha = dp/dd = \gamma$ (Fig. 1.8b; Eq. 1.16b), the patterns of the $p(d)$-profiles and pressure gradients may be summarized as:

$$\tan \alpha_1 = \gamma_{1dyn} < \tan \alpha_2 = \gamma_{2dyn} = \gamma_{st} < \tan \alpha_3 = \gamma_{3dyn}. \qquad (2.4)$$

Note that although the $p(d)$-profiles are treated above as straight lines, they are actually curved. The slopes, $\tan \alpha_1$ and $\tan \alpha_3$ of the lines 1 and 3, and therefore also the vertical pressure gradients γ_1 and γ_3 in the recharge and, respectively, discharge regions, vary with depth. They diverge from the hydrostatic values maximally at shallow depths and approach the static gradient $\tan \alpha_2 = \gamma_{st}$ gradually as depth increases. The curving of the $p(d)$ lines reflects the gradually increasing spacing between the equipotential lines and their tending towards the vertical with increasing depth (Figs. 2.2a,b). A corollary to these changes is that the intensity, or flux, of the vertical component of the flow diminishes gradually with depth and towards the midline area, becoming zero along the basal boundary and the midline.

The vertical pressure gradient γ_{xz} at any given depth, d_{xz}, is represented, therefore, by the slope of the tangent to the $p(d)$-curve at that point of depth:

$$\gamma_{xz\,dyn} = \frac{dp}{dd_{xz}} = \tan \alpha_{xz}. \qquad (2.5)$$

2.2.3 *Dynamic pressure increment:* Δp

The dynamic pressure increment was defined as:

$$\Delta p = p_{dyn} - p_{st} = p_{real} - p_{nom}, \qquad (1.20b)$$

where $p_{st} = \gamma_{st}d$.

Applying this definition to the three numbered sites of the unit basin (Figs. 2.1, 2.2a), the dynamic pressure-increment appears on the $p(d)$ curves for any given depth as, for instance, the difference in pressures $p_1 - p_2 = -\Delta p_1$ between curves 1 and 2, or $p_3 - p_2 = +\Delta p_3$ between curves 3 and 2 (Fig. 2.2b). The dynamic pressure increment for curve 2 is $p_2 - p_2 = \Delta p_2 = 0$ because the flow has no vertical component along the hydraulic midline.

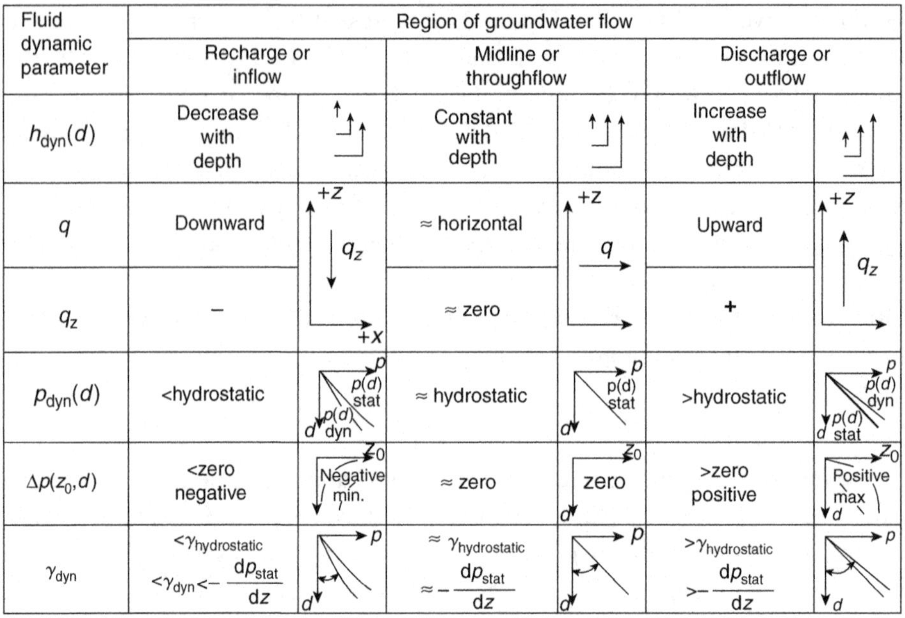

Fluid dynamic parameter	Region of groundwater flow					
	Recharge or inflow		Midline or throughflow		Discharge or outflow	
$h_{dyn}(d)$	Decrease with depth	[diagram]	Constant with depth	[diagram]	Increase with depth	[diagram]
q	Downward	$+z$, q_z [diagram]	≈ horizontal	$+z$, q [diagram]	Upward	$+z$, q_z [diagram]
q_z	−	$+x$ [diagram]	≈ zero	[diagram]	+	[diagram]
$p_{dyn}(d)$	<hydrostatic	p, $p(d)_{stat}$, $p(d)_{dyn}$, d [diagram]	≈ hydrostatic	p, $p(d)_{stat}$, d [diagram]	>hydrostatic	p, $p(d)_{dyn}$, $p(d)_{stat}$, d [diagram]
$\Delta p(z_0,d)$	<zero negative	z_0, Negative min. [diagram]	≈ zero	z_0, zero [diagram]	>zero positive	z_0, Positive max., d [diagram]
γ_{dyn}	<$\gamma_{hydrostatic}$ <γ_{dyn}<$-\dfrac{dp_{stat}}{dz}$	p, d [diagram]	≈ $\gamma_{hydrostatic}$ ≈ $-\dfrac{dp_{stat}}{dz}$	p, d [diagram]	>$\gamma_{hydrostatic}$ >$-\dfrac{dp_{stat}}{dz}$	p, d [diagram]

Fig. 2.3 Diagnostic regional properties of fluid-dynamic parameters in the unit basin.

Whether obtained from $p(d)$-profiles based on observations or calculated numerically, the dynamic pressure increments can be plotted for the entire basin as a function of depth and water-table elevation $\Delta p(d, z_0)$, (Fig. 2.2c; Tóth, 1978, 1979). Such plots show that pressures are subnormal, i.e. lower than hydrostatic, beneath areas above medium basinal elevations. The *deficiency* of pressure increases gradually with depth (note: not the pressure itself!), thereby inducing downward flow. A vertical band of zero-Δp values extends under medium elevations, indicating that only lateral flow is possible at the midline. The distribution of the positive Δp-contours reflects ascending flow in low-lying areas, with increasing intensity towards decreasing water-table elevations and shallower depths.

The salient features of the distribution patterns of the fluid-dynamic parameters in the unit basin are summarized in Figure 2.3.

3

Flow patterns in composite and heterogeneous basins

3.1 Effects of basin geometry

Owing to the functional relations between boundary conditions and hydraulic head distribution in a flow domain (Eqs. 2.1 and 2.2, respectively), changes in the geometry of the unit basin modify the patterns of flow and fluid-dynamic parameters. Typically, the result is an increase in the complexity of the basic distributions. Regional flow patterns are sensitive to two principal aspects of a basin's geometry, namely, the configuration of the water table and the relative depth of the basin, i.e. its depth-to-width ratio (Tóth, 1963).

The analysis below follows Tóth's (1963) original study. First, the linear upper boundary of the unit basin is replaced by an undulating sinusoidal surface superimposed on a linear regional slope. Next, the basin's depth is changed while keeping the width of the basin's flank constant. The effects of significant breaks in the regional slope are then illustrated by conceptual schemata of four major types of regional land forms: (a) V-notch canyons, (b) intracratonic broad upland basins, (c) intermontane broad valleys, (d) cordillera-cum-foreland basins. Finally, Zijl's (1999) analysis of the effect of scale of the water-table's relief on flow-system types and their depth of penetration, and of the relation between spatial and temporal scales is reviewed.

3.1.1 Effect of water-table configuration

The basic effects of the water table's relief on gravity-driven groundwater flow patterns in basins of homogeneous rock framework can be studied by substituting a sinusoidal surface for the linear upper boundary of the unit basin (Tóth, 1962b, 1963, App. B). Recalling that the hydraulic head in a given point of the water table h_{wt} is equal to the water table's elevation at that point (Fig. 2.1), the hydraulic head along the sinusoidal upper boundary can be described as (Figure 3.1):

$$h_{wt} = z_0 + x \tan \alpha + a \frac{\sin(bx/\cos \alpha)}{\cos \alpha}, \tag{3.1a}$$

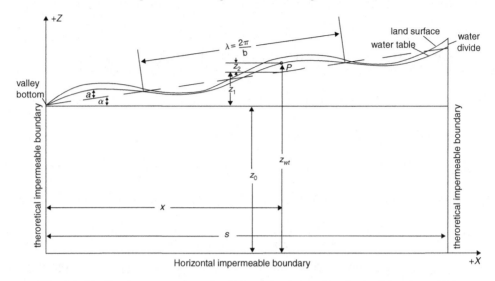

Fig. 3.1 Idealized cross-section of a drainage basin's valley flank of the sinusoidal water table (Tóth, 1963, Fig. 1, p. 4796).

where, z_0 is the elevation of valley bottom above the impermeable base (i.e. basin depth at the thalweg); x: horizontal distance from the thalweg; α is the angle of the regional slope; and a, λ and $b = 2\pi/\lambda$ are the amplitude, wavelength, and wave number, respectively, of the water-table undulation; $x \tan \alpha = z_1$; and $a \frac{\sin(bx/\cos\alpha)}{\cos\alpha} = z_2$ (Fig. 3.1).

Using the substitutions $c' = \tan\alpha$, $a' = a/\cos\alpha$, and $b' = b/\cos\alpha$, with Equation (3.1a), the hydraulic head distribution at the upper boundary is:

$$h_{wt} = z_0 + c'x + a'\sin(b'x). \tag{3.1b}$$

The conditions for the three no-flow boundaries are identical to those established for the unit basin (Eq. 2.1a), namely:

$$\left(\frac{\partial h}{\partial x}\right)_{x=0} = \left(\frac{\partial h}{\partial x}\right)_{x=s} = \left(\frac{\partial h}{\partial z}\right)_{z=0} = 0. \tag{3.1c}$$

The solution of the Laplace equation (Eq. 1.2) with the four boundary conditions (Eqs. 3.1b, c) results in an equation for the hydraulic head at any point P_{xz} in the basin with homogeneous isotropic rock framework and a sinusoidal water table (App. B; Tóth, 1963, p. 4798):

$$h = z_0 + \frac{c's}{2} + \frac{a'}{sb'}(1-\cos(b's)) + 2\sum_{m=1}^{\infty}\left[\frac{a'b'(1-\cos(b's)\cos(m\pi))}{b'^2-m^2\pi^2/s^2}\right.$$
$$\left. + \frac{c's^2}{m^2\pi^2}(\cos(m\pi)-1)\right]\frac{\cos(m\pi x/s)\cosh(m\pi z/s)}{s\cosh(m\pi z_0/s)}, \tag{3.2}$$

where $m = 1, 2, 3, \ldots, \infty$.

If Equation (3.2) is solved for an adequate number of points, P_{xz}, equal values of hydraulic head can be contoured and flow nets constructed. The effects of variations of the basin's shape on the geometric patterns of the flow field and hydraulic-head can then be analysed by comparing flow nets constructed for basins with different values of the parameters of regional slope c', the amplitude of the local topography a, and depth-to-width ratio z_0/s.

3.1.1.1 Effects of undulations of the water table

The primary effect of water-table undulations on basinal water flow is the generation of hierarchically nested *flow systems* of different orders (Fig. 3.2). Strictly defined (Tóth, 1963, p.4806), 'a flow system is a set of flow lines in which any two flow lines adjacent at one point of the flow region remain adjacent through the whole region; they can be intersected anywhere by an uninterrupted surface across which flow takes place in one direction only'. Similar to the unit basin's basic pattern, each individual system has three identifiable segments, or flow regimes, also in the complex basin, namely: recharge, midline and discharge. However, three different orders of flow systems may be distinguished in the complex case: local, intermediate, and regional (Figs. 1.4c, 3.2a).

A system is termed *local* if its recharge and discharge areas are contiguous, *intermediate* if these areas are separated by one or more local systems but do not occupy the main divide or valley bottom and *regional* if it links the basin's principal divide and thalweg hydraulically. The depth of penetration of the various flow systems is a function of the relative magnitudes of the local relief and regional slope. It may exceed a thousand metres in a homogeneous rock framework under the effect of a local relief of a few tens of metres. The changes in spacing between the equipotential lines show that the intensity of the flow decreases, thus the water's residence time increases, from the local to the regional systems (the questions of 'penetration depth' and 'characteristic time' have been analytically investigated by Zijl (1999) and are reviewed later in Section 3.1.3).

Patterns of multiple flow systems are characterized by *alternating regions of recharge and discharge* (Fig. 3.2a). A result of the alternating arrangement is that waters recharged on a given water-table mound may be destined for different discharge areas or, conversely, waters discharging in a given area side by side, or even mixed by diffusion and/or dispersion, may have infiltrated in different parts of the basin.

Another important feature of the complex flow pattern is the possible existence of '*quasi-stagnant' zones* (or, mathematically, '*singular points*') at locations where flow systems of any order may converge from, or part towards, opposite directions. In these areas, the low or zero lateral hydraulic gradients, combined with the

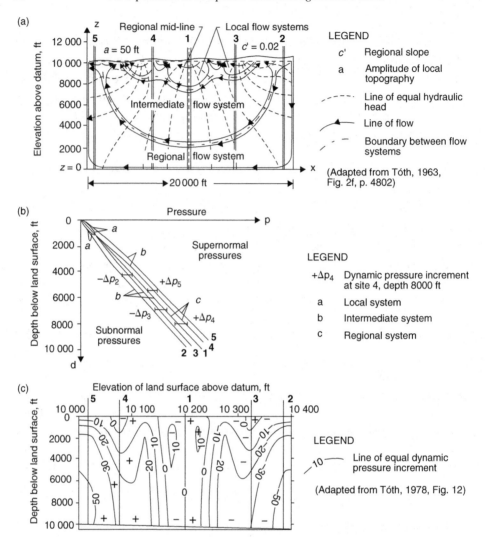

Fig. 3.2 Relation between fluid-dynamic parameters in a basin of complex water-table relief: (a) hydraulic head, $h(x,z)$ and flow, $q(x,z)$; (b) pore pressure, $p(d)$; and (c) dynamic pressure increment, $\Delta p(z_0, d)$ (Tóth, 1980, Fig. 4, p. 126; reprinted by permission of the AAPG, whose permission is required for further use).

corollary poor transport ability of the waters and converging flow directions, have significant ramifications for the accumulation of matter and/or geothermal heat, as exemplified by many anomalously warm fields of petroleum.

Finally, it should be noted that the equipotential lines cross flow-system boundaries without abrupt, or any, changes in direction. Accordingly, no sudden changes in water levels or pore pressures can be expected when crossing flow-system boundaries by boreholes.

The *pressure conditions* that may be expected in complex flow-systems are shown in Figures 3.2(a–c). Again, the vertical pressure-gradient is hydrostatic beneath the regional midline area as indicated by the $p(d)$-profile, and the 0 value for Δp for Site 1 (Figs. 3.2b,c). An inverse relation between the regional water-table elevations and the vertical pressure-gradient is reflected by the counter-clockwise sweep of the $p(d)$-profiles from sites of high altitude (2 and 3) towards the low ones (4 and 5). Sharp changes, indeed trend-reversals in the vertical pressure-gradients are suggested by $p(d)$ curves 3 and 4 crossing over the hydrostatic pressure curve 1 (Figs. 3.2a,b). The depth of the point where the value of the tangent to the local segment of the $p(d)$-curve is hydrostatic is an approximate indication of the depth of the local flow systems. The *regional* recharge- and discharge-character of the areas upslope and downslope from the midline are indicated by the dominantly negative and, respectively, positive Δp-fields in the $\Delta p(z_0, d)$ diagram (Figs. 3.2a,c). The alternating recharge and discharge zones of the shallow local and intermediate systems are reflected, on the other hand, by the alternating signs of the dynamic pressure increments at shallow depths across the elevation range of the water table.

3.1.1.2 *Effects of the regional slope and local relief of the water table*

A general increase in the regional slope, c', of the valley flank results in an increased component of lateral-flow towards the valley bottom. The extent of the surface areas of local systems flowing towards the basin's centre increase, while those of the oppositely oriented systems decrease. The latter systems may degenerate to stagnant regions and even vanish, thus allowing intermediate and regional systems to form (Figs. 3.3a,b).

The interplay between the regional and local relief of the water table can be well observed by comparing the flow patterns in Figures 3.3(b) and (c). Although the steep regional slope of Figure 3.3(b) is maintained, its well developed regional system is replaced by local systems driven by the high local relief and thus reaching the impermeable boundary in Figure 3.3(c). The various flow-pattern combinations can thus be bracketed between two end members: a linearly sloping water table without local undulation, resulting in one single basinal-scale local-system (the unit basin: Fig. 2.1) on the one hand, and a local relief superposed on a horizontal regional surface, with several but only local systems to form (e.g. Fig. 3.3c), on the other.

3.1.2 *Effect of basin depth*

The effect of a basin's depth on the flow pattern depends also on the basin's width. It is more informative, therefore, to analyse this effect in terms of the combined

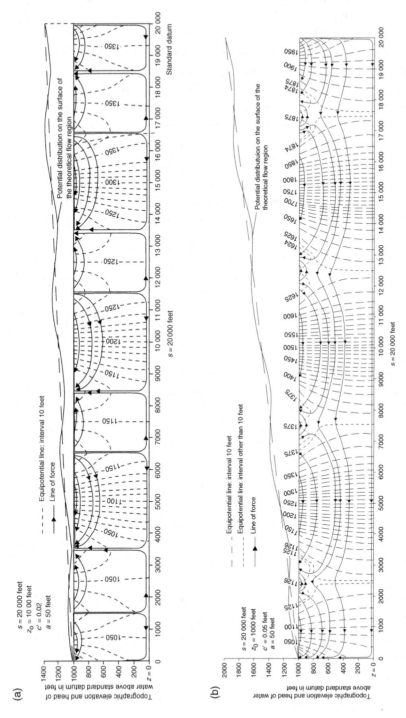

Fig. 3.3 Effect of regional slope, c', amplitude of local water-table relief, a, and thalweg depth, z_0, on the groundwater flow pattern in composite basins: (a) $c' = 0.02$, $a = 50$ ft, $z_0 = 1000$ ft; (b) $c' = 0.05$, $a = 50$ ft, $z_0 = 1000$ ft; (c) $c' = 0.05$, $a = 200$ ft, $z_0 = 1000$ ft; (d) $c' = 0.05$, $a = 200$ ft, $z_0 = 5000$ ft (Tóth, 1963, Figs. 2a, c, d, and e, resp. pp. 4799–4801).

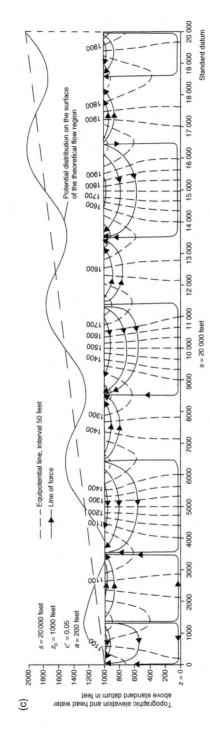

Fig. 3.3 (cont.)

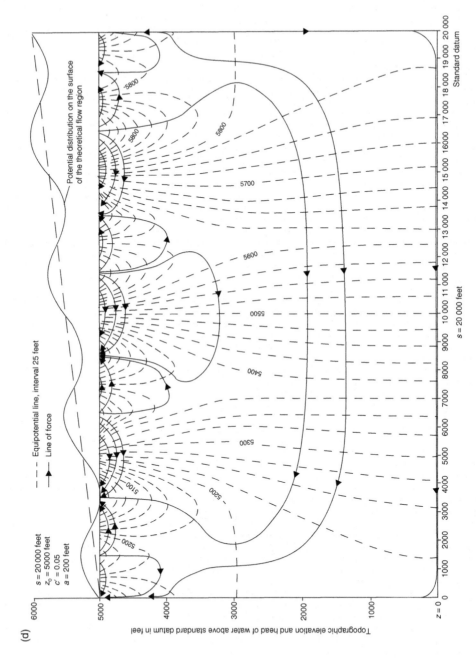

(d)

Fig. 3.3 (cont.)

parameter *relative depth* z_0/s, i.e. the ratio of depth at the thalweg z_0 to the flank's width s, than solely in terms of the depth. An appreciation of the effect of depth can be obtained by comparing the flow pattern of a shallow basin to that of a deeper one. Figures 3.3(c) and (d) illustrate two such basins with depth-to-width ratios z_0/s of 1/20 and, respectively, 1/4. The configuration of the water table is identical in both cases.

The principal difference between the two flow patterns is that only local systems are generated in the shallow basin (Fig. 3.3c), while sufficient room is available in the deeper basin for intermediate and regional systems to form (Fig. 3.3d). Because of the identical water-table configurations, the amounts and distribution of the flow-inducing forces, i.e. the upper boundary conditions, are equal also. Consequently, the same amounts of water are driven through both basins. However, due to the deep basin's larger vertical cross-section, the flow intensity through it is generally less than that in the shallow case. In particular, the local systems of the deep basin reach more deeply thus are less intense than in the shallower one. Several levels of quasi-stagnant zones also may exist in the deeper basin.

3.1.3 Zijl's analysis of the scales of water-table relief, depths of flow-system penetration, and relation between spatial and temporal scales

In a series of three seminal papers Freeze and Witherspoon (1966, 1967, 1968) placed the theory of gravity-driven groundwater flow into the realm of the real world. Based on the equation of hydraulic continuity ($\nabla \cdot \mathbf{q} = 0$) and on Darcy's Law ($\mathbf{q} = -K\nabla\Phi$) they have applied numerical techniques to evaluate the effects of arbitrary topography and heterogeneous and anisotropic rock framework. They have thus liberated the study of flow systems from the strictures of analytical solutions.

The first, and since those papers so far the only step in the theory's conceptual advancement is due to Zijl and Nawalany (1993). They examined and expanded on several aspects of the theory and its possible practical applications by classical mathematical analysis. One such aspect was the often asked but never before answered question of how to rationalize the definition and the naming of flow systems of different hierarchical order. In other words, how to remove the subjectivity from the commonly used qualifiers of 'local', 'intermediate' and 'regional', terms which depend on the size of the area of interest. A corollary question was how to quantify various hydraulic and hydrologic properties of given flow systems, namely: penetration depth; flux and its lateral and vertical variations; the lag time, or characteristic time, between a change in water-table elevation and the response of the flow to it in the associated system; and the relation between the spatial and temporal scales of flow systems. The following review of the scales and hydrologic effects of the water-table relief is based on Zijl's paper (1999) and uses its notation.

The key idea in Zijl's (1999) approach to defining the scales of flow systems independently of the size of the area of interest is to resolve the topographic relief of the water table, $h_f(x, y, t)$, into individual periodic functions, $H_{(x,y)}(\omega, \xi, t)$, called Fourier components or Fourier modes (Zijl,1999, p. 143, Eqs. 3a and b):

$$h_f(x, y, t) = \int_{-\infty}^{\infty} \int_{-\infty}^{\infty} H_{(x,y)}(\omega, \xi, t)\, d\omega\, d\xi \tag{3.3}$$

where

$$H_{(x,y)}(\omega, \xi, t) = A(\omega, \xi, t) \cos(\omega x + \xi y) + B(\omega, \xi, t) \sin(\omega x + \xi y). \tag{3.4}$$

In these expressions $h_f(x, y, t)$ [L] is the hydraulic head at the water table, i.e. the elevation of the water table with reference to the datum plane $z = $ constant, here chosen as $z = 0$; $x[L]$ and $y[L]$ are the co-ordinates along the X and Y axes; $t[T]$ is time; and $H_{(x,y)}(\omega, \xi, t)\, d\omega\, d\xi\,[L]$ is the value of the specific Fourier component for wave numbers $\omega[L^{-1}]$ and $\xi[L^{-1}]$ in the X and, respectively, Y directions. A and B are constants (i.e. independent from X and Y) to be determined from the amplitude, $\sqrt{(A^2 + B^2)}$, and phase, $\arctan(B/A)$, of the specific Fourier component of the water-table elevation. The actual water table elevation is given by $z = h_f(x, y, t)$, where $h_f(x, y, t)$ is considered to consist of the integral of the individual periodic functions, $H_{(x,y)}(\omega, \xi, t)\, d\omega\, d\xi$. Or, as a practical approximation, as the sum of a finite number—in many cases a relatively small number—of individual periodic functions, $H_{(x,y)}(\omega, \xi, t)\, d\omega\, d\xi\,[L]$, that are characterized by their particular wavelength $\lambda = 2\pi \sqrt{(\omega^2 + \xi^2)}$ [L]. (Note that the thus-defined $H_{(x,y)}(\omega, \xi, t)$ is equal to the integral over suitably chosen spectral intervals $\omega_j = (1/\Delta\omega_j) \int \Delta\omega_j\, d\omega$, $\xi_j = (1/\Delta\xi_j) \int \Delta\xi_j\, d\xi, j = 1, 2, \ldots, n$, of the earlier-defined $H_{(x,y)}(\omega, \xi, t)d\omega\, d\xi$.) Flow systems are thus envisaged as generated by gradients of fluid potentials caused by spatial variations in the elevation of the water table. This way the fluid potentials are proportional to the sum of the Fourier components averaged over those intervals and multiplied by the constant of proportionality ρg.

According to this conceptualization, the fluid potential Φ is represented as the projection of the water table onto the horizontal plane $z = 0$: $\Phi(x, y, 0, t) = f(x, y, t) = \rho g h_f(x, y, t)$. Expressed in terms of Fourier components $\Phi(x, y, 0, t) = F_{(x,y)}(\omega, \xi, t) = \rho g H_{(x,y)}(\omega, \xi, t)$.

In this formulation the dimension of the fluid potential is pressure, $[\Phi] = [M/T^2L]$, with $\rho g = 1$ dbar/m where, by definition, 1 dbar $= 10^4$ kg/s^{2}m. A pressure of 1 dbar thus corresponds to a column of fresh water with a height of 1 m. If this equality is assumed, as is customary in hydrogeological practice, then

$f(x, y, t) = h_f(x, y, t)$. Use of the right-hand side expression, however, does not reveal the role of gravity and density in the mathematical formulations. In order, therefore, to underscore that role in dealing with a problem of gravity-driven flow, Zijl (1999) has developed the following equations with ρg shown explicitly.

With the fluid potential defined, the fluid-driving force, and hence the horizontal and vertical flux, can be calculated. To simplify considerations, the horizontal co-ordinate axes are rotated into an X'–Y' co-ordinate system so that the fluid potential can be written as a function only of x' and t:

$$F_{(x')}(\lambda, t) = \rho g \left\{ A(\lambda, t) \cos(\omega' x') + B(\lambda, t) \sin(\omega' x') \right\}, \tag{3.5}$$

where $\omega' = \sqrt{(\omega^2 + \xi^2)} = 2\pi/\lambda$.

After differentiation of Equation (3.5) (to determine the driving force from the potential) and applying Darcy's Law (to determine the horizontal fluxes caused by the Fourier component considered) the horizontal flux for an infinitely deep basin is obtained:

$$Q_{X'(x')}(\lambda, 0, t) = \rho g \omega' K_h \left\{ A(\lambda, t) \sin(\omega' x') - B(\lambda, t) \cos(\omega' x') \right\}. \tag{3.6}$$

After differentiation of Equation (3.6) and applying the continuity equation and Darcy's Law (to determine the vertical fluxes caused by the Fourier component considered) the vertical flux for an infinitely deep basin is obtained:

$$Q_{Z'(x')}(\lambda, 0, t) = \rho g \omega' \sqrt{(K_h K_v)} \left\{ A(\lambda, t) \cos(\omega' x') + B(\lambda, t) \sin(\omega' x') \right\}. \tag{3.7}$$

The magnitude of the flux components at depth z is (Zijl, 1999, p. 143, Eqs. 8a and 8b):

$$Q_{X'(x')}(\lambda, z, t) = Q_{X'(x')}(\lambda, 0, t) \exp\left\{ -(2\pi z/\lambda) \sqrt{(K_h/K_v)} \right\}, \tag{3.8a}$$

$$Q_{Z'(x')}(\lambda, z, t) = Q_{Z'(x')}(\lambda, 0, t) \exp\left\{ -(2\pi z/\lambda) \sqrt{(K_h/K_v)} \right\}, \tag{3.8b}$$

where $\beta = (2\pi z/\lambda) \sqrt{(K_h/K_v)}$ is the exponential rate of decay of the flux's intensity with depth. This rate can be appreciated by comparing the factors of flux-intensity decay, $\exp(-\beta)$, determined for different depths. If these depths are chosen as fractions of the wavelength λ as, for instance, $z = \lambda/4, \lambda/2, 3\lambda/4, \lambda$, then, in a homogeneous and isotropic rock framework, the corresponding values of β and the factors of decay, $\exp(-\beta)$, are, respectively, as follows: $\beta = \pi/2, \pi$, $3\pi/2$ and 2π and $\exp(-\beta) \approx 0.21, 0.043, 0.01$ and 0.002. It seems therefore that, somewhat depending on the purpose of the analysis, the depth of an effectively impermeable base, or penetration depth, can be quantitatively estimated based on the factor of decay. If, for instance, a 0.2 per cent loss from the flow system's flux is acceptable for the objective of the analysis, then the depth z at which the factor of decay, $\exp(-\beta)$, is 0.002 (i.e. $\beta = 2\pi$ and $z = \lambda$), may be deemed as 'effectively impermeable' for the flow system of wavelength λ. Thus from Equations (3.8a) and (3.8b) the penetration depth, δ, is defined as (Zijl, 1999, p. 143, Eq. 9):

$$\delta = \lambda\sqrt{(K_v/K_h)}. \tag{3.9}$$

The choice of $\beta = 2\pi$, i.e. $\delta = z = \lambda$, is not entirely arbitrary. Rather, it is based on the fact that a hydraulic conductivity distribution different from K_v and K_h below $z = \delta$ has little effect on the strength of flow, whereas such a disturbance above it has a considerable influence (Meekes, 1997). Figure 3.4(a) illustrates the effectively impermeable shallow base, or penetration depth, of flow systems generated by short wavelength undulations of the water table, while Figure 3.4(b) shows the deeper reach of systems induced by longer waves. 'Since the shallow flow velocities caused by the short waves have much larger flux intensity than the shallow flow velocities caused by the long waves, the shallow streamlines are almost fully determined by the short wavelengths. On the other hand, at depths below the effectively impervious base of the short waves, the flow velocities caused by the long waves are dominant. Hence the deep streamlines are almost fully determined by the long waves.' (Zijl, 1999, p. 144).

The hierarchical nesting of flow systems of three different orders generated by three different wavelength components of the water table (short or 'local', 'intermediate' and long or 'regional') is illustrated in Figure 3.5. The dashed lines indicate boundaries between the domains of the different orders of flow systems with quasi-stagnant regions (*singular points*) located at their nodes.

Unlike in a homogeneous rock framework, the depth of the 'effectively impervious base' is sharply defined in the cases of horizontally layered alternating aquifers and aquitards. Instead of the exponentially diminishing smooth decay of flux intensity with increasing depth of the first case, a stepwise change occurs across the boundary between the base of an aquifer and the top of the subjacent aquitard.

Zijl (1999) discussed quantitative relations also between the spatial and temporal scales of groundwater flow systems. He used wave length to characterize the relief of the water table, and characteristic time to measure flow-system sensitivity to show that 'the dependence [on time] of the amplitude of a Fourier mode depends on the wave length λ of that component' (Zijl, 1999, p. 144). His analysis starts with the kinematical boundary condition on the water table, which holds for each Fourier component separately:

$$\frac{n\partial H_{(x')}(\lambda, t)}{\partial t} = P_{(x')}(\lambda, t) - Q_{Z(x')}(\lambda, 0, t), \tag{3.10}$$

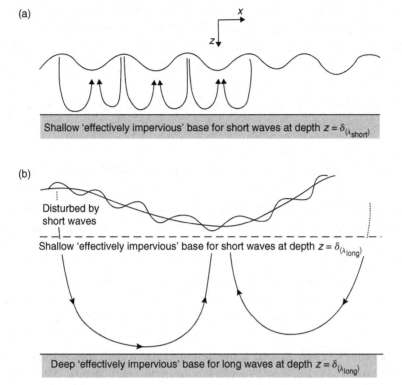

Fig. 3.4 Penetration depths considered as 'effectively impervious' bases for (a) short and (b) long waves in the water table and the resulting streamline patterns (Zijl 1999, Fig. 1, p. 144).

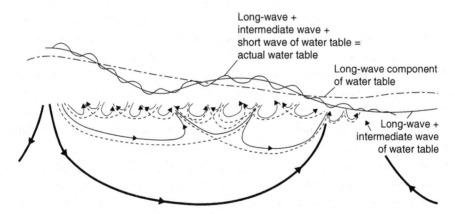

Fig. 3.5 Superposition of long, intermediate and short waves of the water table and the resulting streamline pattern (Zijl, 1999, Fig. 2, p. 144).

where n is the specific yield of the zone of the water table ($\approx$ porosity); $H_{(x')}(\lambda, t)$ is the Fourier component of the water table along co-ordinate axis X'; $P_{(x')}(\lambda, t)$ is the Fourier component of that part of precipitation replenishing the groundwater body; and $Q_{Z(x')}$ is the vertical flux of descending groundwater flow projected onto the datum plane $z = 0$.

Evaluating $Q_{Z(x')}(\lambda, 0, t)$ from Equations (3.5) and (3.7) and combining with Equation (3.10) yields:

$$\frac{n \partial H_{(x')}(\lambda, t)}{\partial t} = P_{(x')}(\lambda, t) - (2\pi/\lambda)\sqrt{(K_h K_v)}\rho g H_{(x')}(\lambda, t). \tag{3.11}$$

If accretion of the groundwater body from precipitation is assumed to be constant, i.e. $P_{(x')}(\lambda, t) = P_{(x')}(\lambda)$ then the relief of the water table will attenuate asymptotically to a steady state:

$$H_{(x')}(\lambda, t) = H_{(x')}(\lambda, \infty) + \left\{ H_{(x')}(\lambda, 0) - H_{(x')}(\lambda, \infty) \right\}$$
$$\exp\left[-\left\{ (2\pi/\lambda)\sqrt{(K_h K_v)}\rho g/n \right\} t \right]. \tag{3.12}$$

In Equation (3.12) $H_{(x')}(\lambda, 0)$ is the height of the water table at time $t = 0$ and $H_{(x')}(\lambda, \infty) = P_{(x')}(\lambda)/(2\pi/\lambda)\sqrt{(K_h K_v)}\rho g$ is the final, steady-state position of the water table. The characteristic time, which is defined as the time in which the water table has changed from initial elevation $H_{(x')}(\lambda, a)$ to height $H_{(x')}(\lambda, \infty) + \{H_{(x')}(\lambda, t) - H_{(x')}(\lambda, \infty)\}/e$ is then (Zijl, 1999, Eq. 14, p. 145):

$$\tau = n \cdot \lambda / \left\{ 2\pi \sqrt{(K_h \cdot K_v)} \cdot \rho \cdot g \right\}. \tag{3.13}$$

Equation (3.13) shows that the characteristic time increases with increasing wavelength of the water table's undulation. For instance, for a wavelength of 12.6 km, effective hydraulic conductivities $K_h = 1$ m^2/dbar per day and $K_v = 0.0025$ m^2/dbar per day, porosity $n = 0.25$, and $\rho \cdot g = 1$ dbar/m, the characteristic time is $\tau = 10\,000$ days ≈ 27 years. Below the calculated penetration depth of $\delta = \lambda\sqrt{(K_v/K_h)} = 12\,600\sqrt{0.0025} = 630$ m, the flow is not affected by short-term precipitation events such as daily rainfall or even annual changes in weather pattern.

On the other hand, in the case of a local wavelength of 63 m, hydraulic conductivities of $K_h = 1$ m^2/dbar per day and $K_v = 0.1$ m^2/dbar per day, porosity $n = 0.35$, and $\rho g = 1$ dbar/m, the characteristic time becomes $\tau = 11$ days. With penetration depth δ of approximately 20 m, this flow system is sensitive to even weekly changes in precipitation.

3.1.4 Effects of major regional land-form types

In order to highlight and summarize the regional effects of major land forms on basinal flow patterns, schematic distributions are shown for four common topographic settings: (a) V-notch canyon; (b) intracratonic broad upland; (c) intermontane broad valley; and (d) cordillera-cum-foreland (Figs. 3.6a–d). Actually, the first three of these land forms may be considered as unit basins modified by major single breaks

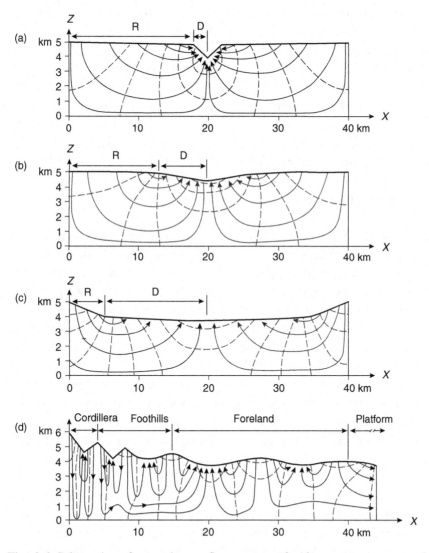

Fig. 3.6 Schematics of groundwater flow patterns for four common types of regional landforms: (a) V-notch canyon; (b) intracratonic basin with broad uplands; (c) intermontane broad valley; (d) cordillera-cum-foreland.

in the water table. The basins are assumed to be approximately 40 km wide and 5–6 km deep. Although these dimensions are arbitrarily chosen, numerous real-life situations can be characterized by them. Positing homogeneity of permeability over rock volumes of such sizes is less realistic. However, in gravity regimes the relief of the water table generates the flow thus controlling its sense and pattern type. Permeability plays a modifying role. Consequently, the principal attributes of flow patterns in basins of given geometric configuration will be preserved under a great variety of permeability distributions.

(a) V-notch canyon basins. A V-notch canyon can be considered as an extreme type of regional land form. It is composed of an extensive and gently sloping flat upland, and a deeply incised, narrow central valley (Fig. 3.5a). Hydraulically, the valley is a line sink: it is the locus of discharge of all waters moving through the basin. On either side of the valley, there is one single area of recharge, R, and one of discharge, D. The recharge areas are much greater than the discharge areas, $R/D \gg 1$ Consequently, flow through most of the basin is descending and/or lateral. Formation-fluid pressures and vertical pressure-gradients are, therefore, largely subhydrostatic and/or hydrostatic, respectively. The quasi-stagnant zone beneath the discharge areas is relatively small, resulting in only a narrow band of superhydrostatic pressures and pressure gradients.

(b) Intracratonic broad upland basins. An intracratonic broad upland basin consists of two relatively flat flanks sloping gently toward a broad and shallow central depression that may be occupied by one or more shallow lakes, meandering rivers with broad flood plains, marshes or some other types of poorly integrated drainage systems (Fig. 3.6b). The regional flow pattern developed under these topographic conditions resembles closely that of the unit basin. Local flow systems are insignificant on the basin's scale and the recharge area is only slightly greater than, or is equal to, the area of discharge, $R/D \geq 1$. Pore pressures and vertical pressure-gradients are hydrostatic or slightly subhydrostatic under wide areas in the midline regions. However, superhydrostatic pressures and pressure gradients, possibly to the degree of flowing-well conditions, also exist under relatively broad regions, indicating a large areal extent for the quasi-stagnant zone of the central discharge area.

(c)Intermontane broad valleys. Intermontane broad valleys are characterized by upward concave and relatively flat river valleys or lake basins bordered by abruptly rising sub parallel or circular mountain ranges (Fig. 3.6c). Formation waters in these basins are recharged primarily in the mountains. Most of the lowland area is, therefore, one of groundwater discharge although some shallow local systems

may develop in response to local undulations of the water table. The ratio of recharge- to discharge-area is $R/D <$ or $\ll 1$. Accordingly, pore pressures and pressure gradients are superhydrostatic under most of the basin's area, possibly reflected by widespread, but mild, flowing-well conditions. The areal extent of the quasi-stagnant central zone is significant as compared to the basin's dimensions. In case of unequal elevations of the bordering mountain ranges, the zone may be shifted towards the side of the lower mountains.

(d) Cordillera-cum-foreland basins. Cordillera-cum-foreland basins comprise one side of cordillera-type high mountain ranges, their foothills, and the adjacent, rolling, plains, i.e. the foreland (Fig. 3.6d). Usually, the mountains are tectonically detached from the foothills by a belt of low angle thrust faults.

It is generally unknown whether the thrust belt is a conduit or barrier to fluid flow between the mountains and the foothills. Similarly open is the question of hydraulic communication between the foothills and the foreland plains. Yet, these questions can be important for certain practical reasons, such as exploration for water resources or petroleum. For instance, the migration trajectories of petroleum transporting fluids, thus locations of potential entrapment sites, are fundamentally different in the foothills and foreland regions if, on the one hand, they are affected by the high fluid potentials of the contiguous physiographic units or, on the other, if they are not. In my opinion, based on theoretical considerations and field data from the Western Canadian Sedimentary Basin (Tóth, 1978) and the Carpathian Basin of Central Europe (Tóth and Almási, 2001), subsurface hydraulic conditions in the high mountains may have some influence on formation-water flow in the foothills, but they have little or no effect in the foreland or beyond.

The theoretical basis of this postulate is the assumption that, due to the great depth and relatively small width of cordilleran valleys (the short wavelength high-frequency 'local' undulations at the mountain-range scale; Fig. 3.5; Eq. 3.9), the penetration depth of the water recharged high in the mountains is relatively shallow. Most of the water recharged there is also discharged locally through the lower slopes of the valleys and the valley floors. Consequently, no or only minor amounts of water are left for deeper infiltration: the high fluid potentials available at the mountain peaks are, therefore, not transmitted far below the valley bottoms. The strongly accentuated topography favours the development of 'local' systems and robs the deep subsurface of regional recharge. Furthermore, the elevation of the valley bottoms is generally not significantly higher than the average elevation of the forelands. The situation results in relatively low regional hydraulic gradients and lets the effect of the local relief prevail (Figs. 3.3b and c). Consistent with this view is the observation that, for instance, in the Rocky Mountain Foothills and forelands in Alberta, Canada, as well as in the Carpathian Basin, no formation pressures and flow patterns are observed that would reflect the transfer of high fluid-potentials from the mountains to the foreland regions [the high formation pressures in the Hungarian Great Plain of the Carpathian Basin have been attributed to tectonic compression by Tóth and Almási (2001)].

Figure 3.6(d) is intended to suggest, therefore, the following principal features perceived to characterize the subsurface flow regimes of cordillera-cum-foreland basins:

(i) In the cordilleras, local flow systems of single, intensive, and deep cells originate in the high mountains and terminate in the valleys. Some very sluggish intermediate systems discharging in the foothills may develop, but no water infiltrating in the mountains reaches the foreland.

(ii) Intensive, deep, local systems, adapted to the major elements of the topographic relief, prevail in the foothills. Nevertheless, due to the less accentuated topography resulting in a reduced ratio of the local relief to the regional slopes, intermediate systems discharging in the foreland are enhanced: flow toward the foreland may take place beneath extensive quasi-stagnant zones associated with broad valleys in the foothills.

(iii) The flow pattern in the foreland is dominated by large local systems adjusted to the principal topographic features of this region. Inter-valley flow (intermediate or regional systems) is possible but rare; its development depends on the ratio of local vs. regional relief and the basin's depth. Foreland basins nearest to the mountains may receive discharge from the foothills. Otherwise, formation-water flow in the foreland is maintained by local recharge. The relative extents of the recharge and discharge areas depend on the morphology of the particular area considered but a ratio of $R/D \approx 1$ may be typical. The discharge areas are associated with the quasi-stagnant zones of the flow systems' ascending limbs and are thus located primarily in the central parts of the valleys of the foothills and the foreland.

3.2 Effects of basin geology

Macro-scale variations, or heterogeneity, in the rock framework's permeability is a third major factor affecting basinal groundwater flow patterns (in addition to water-table relief and basinal depth). A comprehensive overview of the effects of heterogeneity is impractical if not impossible, however, due to the great diversity of its possible types and patterns. Instead, an attempt is made to identify permeability variations that either change the character of a flow field, or are important in some practical hydrogeological problems, such as water-well field design, petroleum exploration, slope stabilization, subsurface contamination, and so on. The rock configurations selected by these criteria are: (1) stratification; (2) lenses; (3) faults and (4) anisotropy.

The effects of permeability variation are studied here by analyzing the differences between *reference patterns* of flow and dynamic parameters calculated for homogeneous 'theoretical' flow domains, on the one hand, and those obtained for heterogeneous 'real-life' domains of similar boundary conditions, on the other.

3.2.1 Effects of stratification

Reference cases for homogeneous basins are shown in Figures 2.2 and 3.7(a), for relatively simple topography, while Figures 3.2 and 3.7(b) represent basins with undulating water tables (Fig. 2.2 and 3.2: Tóth, 1980; Figs. 3.7a and 3.7b: Freeze and Witherspoon, 1967). The Freeze and Witherspoon examples are calculated for dimensionless basins. In the latter cases the basin's depth is normalized to its width s with the depth-to-width ratio being approximately $z_0/s \approx 0.08$.

3.2.1.1 Two-layer cases

(i) Aquifer above aquitard. In Figure 3.8 a high permeability stratum, $k = 100$, overlies an aquitard of $k = 1$. Comparison with Figure 3.7(a) shows little difference between the two patterns of hydraulic head. Yet, flow intensity is one-hundred times greater in the aquifer than in the aquitard as seen from the spacing of the equipotential lines: the spacing, i.e. the driving force, remains the same in both

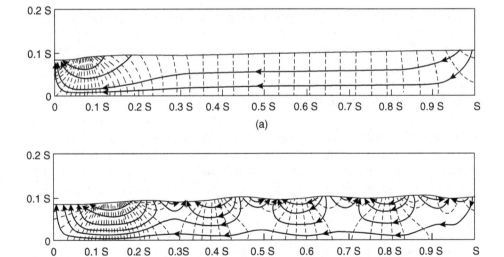

Fig. 3.7 Effect of topography on regional groundwater flow patterns: (a) linear basin surface, broad central valley; (b) undulating basin surface, broad central valley (after Freeze and Witherspoon, 1967, Fig. 1A,C, p. 625.

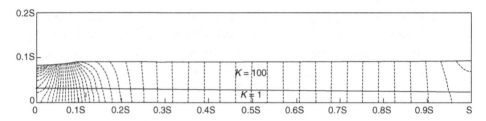

Fig. 3.8 Effect of stratification, aquifer above aquitard (after Freeze and Wither-spoon, 1967, Fig. 2E, p. 627)

strata, but the top one conducts water a hundred times better than the lower one. The base of the domain of strong flow has risen to the bottom of the highly permeable layer. Nevertheless, flow does occur through the aquitard too. The fact that water crosses its upper boundary is indicated by equipotential lines meeting that boundary at angles different from 90° at the recharge and discharge ends of the basin.

(ii) Aquifer below aquitard. Significant effects on the distribution of basinal flow can be caused by highly permeable strata that underlie surficial aquitards. With respect to the patterns of flow and hydraulic-head in a homogeneous basin with no local relief (Fig. 3.7a), the following modifications may be attributed to horizontal aquifers of permeabilities 10 and 100 times greater than that of the overlying surficial aquitard (Figs. 3.9a and b; Freeze and Witherspoon, 1967):

(1) The direction of the descending and ascending flow through the upper portion of the basin (the aquitard in Figs. 3.9a and b) is oriented more towards the vertical as the permeability of the underlying aquifer increases: essentially, the hydraulic polarization between recharge and discharge is enhanced. The effect stops increasing at a permeability ratio of approximately 1000.
(2) With increasing aquifer permeabilities, the hydraulic midline migrates upslope from the area of upward flow toward that of downward flow, thus increasing the discharge area and the lateral extent of the quasi-stagnant zone beneath it.
(3) The greater the permeability contrast between the aquifer and aquitard, the larger is the portion of the total basinal water flow through the aquifer. However, this effect also stops increasing beyond a limiting ratio of permeabilities: at such a ratio the aquifer's capacity to transmit water laterally becomes greater than that of the aquitard for vertical flow.

If the basin is bounded at the surface by an undulating water table, a highly permeable basal aquifer causes a reduction in the extent of the local discharge areas and an expansion of the local recharge areas, in addition to the above mentioned effects (Figs. 3.7b and 3.9c). In extreme situations, the basinal flow pattern of an undulating water table may revert to one of a simple basin.

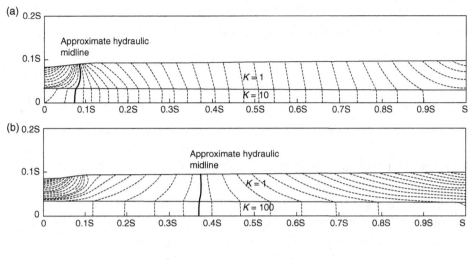

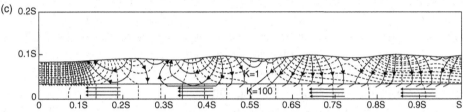

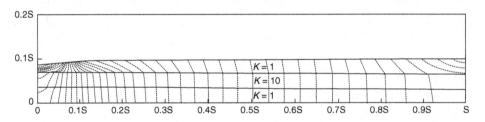

Fig. 3.9 Effect of stratification, aquitard K_1 above aquifer K_2: linear basin surface, broad central valley, (a) $K_1/K_2 = 1/10$; (b) $K_1/K_2 = 1/100$ and (c) undulating basin surface, broad central valley, $K_1/K_2 = 1/100$ (modified from Freeze and Witherspoon, 1967, Figs. 2A, B, p. 627 and 3A, p. 628).

Fig. 3.10 Aquifer sandwiched between two aquitards (after Freeze and Witherspoon, 1967, Fig. 2F, p. 627).

3.2.1.2 Three-layer cases.

(i) Aquifer sandwiched between two aquitards. The general characteristics of this three-layer pattern (Fig. 3.10) can be derived from the combination of the two-layer cases discussed above: increased verticality and intensity of the downward and upward flow through the top aquitard; increased flux through the aquifer; expansion

of the area of discharge and little change in the pattern, but significant reduction in the intensity, of the flow in the bottom aquitard.

(ii) Aquitard sandwiched between two aquifers. A reversed situation is presented in Figures 3.11(a–c), namely that of a slightly permeable, $K_2 = 1$, aquitard separating an upper aquifer of $K_1 = 100$ from a bottom aquifer of $K_3 = 1000$. The patterns of flow and dynamic parameters were generated originally to study general characteristics of the flow field, and to analyse pore–water dynamics in an actual basin, in northern Alberta, Canada (Tóth, 1978, 1979). Corresponding patterns of flow- and fluid dynamic parameters for a unit basin are given in Figure 2.2, which can be used as a reference base in evaluating the effects of the aquitard.

The general characteristics of parameter distribution in the three-layer case appear to be similar to those of the homogeneous basins (Figs. 2.1 and 2.2). However, the areas of recharge and discharge have now shrunk to the vicinity of the divide and thalweg, and flow has become dominantly subhorizontal in the depth ranges of the two aquifers. The changes can be attributed to the reduced depth-to-width ratio in the upper aquifer relative to the homogeneous case, to the high permeability of the lower aquifer, and to the high vertical hydraulic resistance (thickness/permeability) of the aquitard with the corollary requirement of a large surface area of flow through it. Consequently, the area of vertical flow is more extensive in the aquitard than in the aquifers. This situation can be seen from the refraction of equipotential lines (Fig. 3.11a), the high vertical pressure gradients through the aquitard (stippled sections, Fig. 3.11b), and from the high absolute values of the dynamic pressure-increments and an increase in density of their iso-lines (Fig. 3.11c).

Attention is drawn at this point to the possible misinterpretation of actual observations made in similar real-life situations. The characteristic elements of such observations are the following (Figs. 3.11a–c): an absence of pressure measurements through the aquitard, i.e. between the two aquifers; on average, 'normal' hydrostatic pressure gradients in the upper aquifer; and major pressure anomalies in the lower one. The conventional, and erroneous, interpretation of these observations would probably be that:

(a) the upper and lower aquifers of the basin are hydraulically disconnected by an impermeable formation between them;
(b) water is unconfined and static in the upper aquifer; but
(c) it is confined in the lower aquifer, with strong positive and negative pressure anomalies indicating either flow or compartmentalized static conditions.

In fact, all the above phenomena represent the effect of an areally extensive aquitard through which flow takes place between two highly permeable aquifers: the three strata constitute one single, hydraulically continuous flow domain.

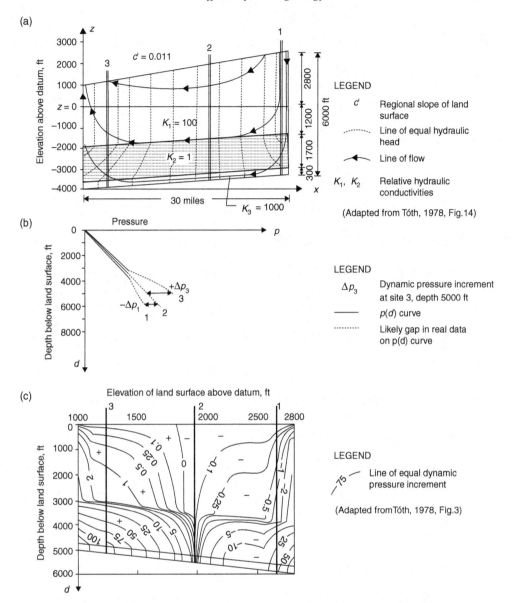

Fig. 3.11 Aquitard sandwiched between two aquifers, effects on: (a) hydraulic heads, $h(x, z)$ and flow, $q(x, z)$; (b) pores pressures, $p(d)$, and (c) dynamic pressure increments, $\Delta p(z', d)$; (after Tóth, 1980, Fig. 3, p. 125; reprinted by permission of the AAPG, whose permission is required for further use).

The effect of an aquitard sandwiched between two aquifers is still further accentuated if the basal aquifer outcrops at an elevation below the base of the aquitard.(A common situation in mountainous and hilly areas.) The dynamic parameter patterns of a relatively general situation (Figs. 3.12a–c) simulate conditions in the

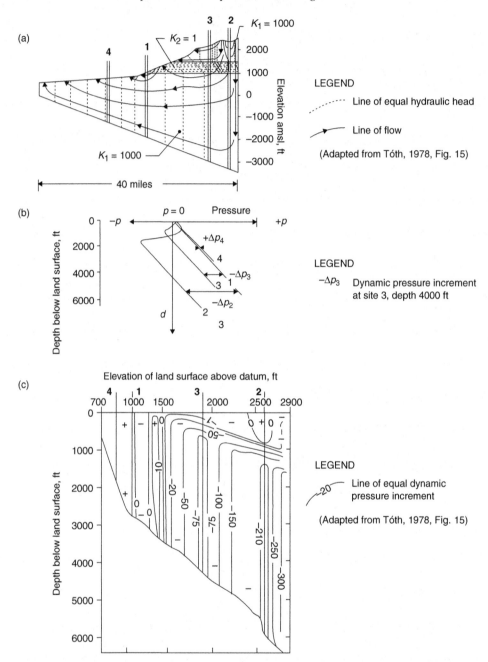

Fig. 3.12 Aquitard sandwiched between two aquifers, with basal aquifer outcropping at low elevation, effects on: (a) hydraulic heads, $h(x, z)$ and flow, $q(x, z)$; (b) pores pressures, $p(d)$, and (c) dynamic pressure-increments, $\Delta p(z', d)$ (after Tóth, 1980, Fig. 5, p. 128; reprinted by permission of the AAPG, whose permission is required for further use).

Late Devonian and younger sediments in northern Alberta, Canada (Tóth, 1978, 1979).

The patterns exhibit most major features of previously discussed cases. Thus, the potential- and flow-distribution is sensitive to the local relief above the aquitard. Consequently, local recharge and discharge areas alternate across the basin as reflected, for instance, by descending flow lines, subhydrostatic vertical pressure gradients, and negative dynamic pressure increments at Site 3, on the one hand, and ascending flow, superhydrostatic pressure gradients and a shallow local field of positive $\Delta p's$ at Site 2, on the other (Figs. 3.12a–c).

These observations show also that most of the flow is deflected along the top of the aquitard and only a fraction of the total recharge enters the unit. The large losses of energy and the relatively small flow through this zone are indicated by a congestion of the lines of equal head (Fig. 3.12a) and by correspondingly rapid drops in pressure and dynamic pressure-increments (Figs. 3.12b,c). Indeed, the deficiency of recharge from above into the lower aquifer, as compared with the aquifer's capacity to transmit lateral flow, can be sufficiently large to generate pore pressures that are negative with respect to atmospheric pressure. If the elevation of the discharge area is low enough to maintain an excess of outflow over replenishment available through the aquitard above suction, or a 'tensiometric effect', may develop in the aquifer some distance away from its outcrop (Tóth, 1981). Air may thus enter into the region at and near the aquifer's edge resulting in the development of a water table. Flow is slow and essentially subhorizontal in the lower zone of high permeability. This is indicated by the wide spacing between and vertical trend of the $h = $ constant equipotentials and iso-Δp lines (Figs. 3.12a,c), and by the essentially hydrostatic rates of pressure increase at Sites 1–3 (Fig. 3.12b). Slightly supernormal rates of pressure increase at Site 4 (Fig. 3.12b) and positive values of the pressure increments at elevations below the outcrop of the aquitard (Figs. 3.12a,c) are indicative of a discharge regime at the outcrop regions of the lower aquifer.

3.2.1.3 Sloping beds outcropping at the land surface

Aquifers and aquitards that are exposed at the land surface may affect the groundwater flow conditions in the vicinity of the outcrop area. The effects depend on a large number of factors, such as: the hydraulic regime of the area (recharge, midline, discharge); hydrostratigraphic nature (aquifer, aquitard), orientation, dip and thickness of the outcropping stratum; and permeability contrast with the surrounding rocks. The main forms of the effects include: reversal of recharge or discharge conditions, modification of flow intensity and changes in the areal extents of recharge and discharge regions. These effects may be particularly important in the hydrogeological interpretation of surface geochemical phenomena (to be discussed later).

The only example presented here is from Freeze and Witherspoon (1967) and it is intended to dispel a time honored fallacy in some geologists' thinking. In

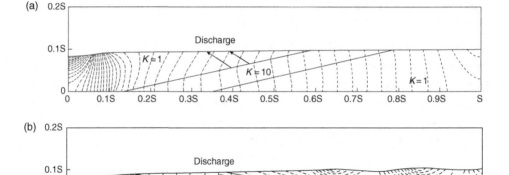

Fig. 3.13 Regional groundwater flow in sloping aquifers outcropping in discharge area: (a) outcrop enhancing recharge; (b) outcrop causing discharge (modified from Freeze and Witherspoon, 1967, Fig. 5A, C, p. 630).

that view, the outcrop of a highly permeable bed is synonymous with an area of groundwater recharge. A permeable outcrop is thus axiomatically accepted to induce groundwater recharge, i.e. down-dip flow. Recharge is enhanced, indeed, by the aquifer in Figure 3.13(a). However, this is so only because this outcrop is located in an area that would be one of recharge in a homogeneous basin also. At the same time, part of the induced water is discharged immediately downslope from the point where the aquifer's upper boundary intersects the water table; a would-be area of recharge without the presence of the aquifer. On the other hand, the aquifer in Figure 3.13(b) causes groundwater to discharge by outcropping in an area that would be one of recharge in a homogeneous basin. Clearly, the mere presence of a highly permeable rock at the surface is not a sufficient condition for down-dip flow.

3.2.2 Effects of lenses

Highly permeable lenticular rock bodies can cause significant perturbations in the groundwater flow and fluid-potential fields, as shown by theoretical and field studies (Tóth, 1962a, 1966b; Freeze and Witherspoon, 1967; Obdam and Veiling, 1987; Tóth and Rakhit, 1988; Parks, 1989; Fitts, 1991). Such perturbations can encompass portions of the basin that are considerably larger than the causative rock pod and may alter basinal aspects of groundwater such as, for instance, its hydrology, chemistry, temperature and surface manifestations. Modifications of the flow field of such areal extent are termed here the *basin-scale effects* of lenticular rock bodies. Another type, the *lens-scale effects,* are limited to distances around the rock lens comparable to the lens' own dimensions. In applying hydrogeology to petroleum exploration, for instance, basin-scale effects should be considered when planning and/or interpreting surface geochemical measurements (to be discussed in Chapter 5). Lens-scale effects, on the other hand, may serve as direct potentiometric indicators of the presence of reservoir quality rock.

3.2.2.1 Basin-scale effects of lenticular rock bodies

One of the most significant basin-scale effects of lenticular rock bodies is the possible change in, or even reversal of, the type of groundwater flow-regime between the lens and the land surface. Other effects include modifications in the hydrologic balance of the land surface and radical changes in flow-path trajectories (possibly important factor in contaminant migration).

By means of a series of computed flow fields, Freeze and Witherspoon (1967) showed that areas of lateral flow and recharge in a homogeneous basin can be converted into areas of discharge by a highly permeable lens located beneath the upslope regions (compare Figs. 3.7a and 3.14). The effect is the opposite if the lens is located in the downslope part of the basin: areas of lateral and ascending flow of a homogeneous basin may thus be changed into regions of descending flow. In both cases, the effects may be intensive enough to cause the segmentation of a hydraulically simple basin (with one of each of recharge, midline, and discharge area) into a region of multiple recharge, midline, and discharge areas. The appearance of several segregated areas of the same hydraulic type in a topographically featureless basin may be puzzling without the understanding of the possible effects of heterogeneities on the flow field. On the other hand, such understanding may lead to useful inferences on the subsurface geology and groundwater flow directions from surface observations.

3.2.2.2 Lens-scale effects of lenticular rock bodies

Lens-scale effects are perturbations in the potentiometric and flow fields comparable in size to the causative rock bodies encased in a rock matrix of different permeability. In nature, a rock pod may have greater or lower permeability than that of its surroundings. Pods of high permeability are of particular interest because

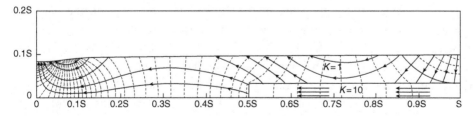

Fig. 3.14 Groundwater flow field modified by highly permeable lens located beneath the upslope region of a basin with linearly sloping surface and broad central valley (after Freeze and Witherspoon, 1967, Fig. 4A, p. 629).

they can concentrate flow and entrap transported matter, such as contaminants, hydrothermal fluids carrying salts of metals, or natural hydrocarbons. The present discussion is focused, therefore, on the effects of relatively highly permeable rock lenses.

The perturbations in the potentiometric field depend on a number of parameters related to the geometry, permeability, and orientation of the rock body. The relations may be studied by calculating the effects on the fluid-potential field of rock lenses whose various parameters are systematically changed. Figure 3.15 shows an elliptical rock lens of width W, length L, and permeability, k', placed in a rock matrix of homogeneous permeability, k, and in an originally uniform flow field of $\mathbf{q}_o$ and Φ_o. The presence of the lens induces the perturbed, or anomalous, fields of flow, $\mathbf{q}_a$, and fluid potential, Φ_a.

As a basic measure of the lens's effect, the absolute anomaly (or simply anomaly) of the fluid potential $\Delta\Phi$, or hydraulic head Δh, is defined as the difference between the perturbed and the original, or unperturbed, potential or hydraulic-head fields at a given point in the flow domain (Fig. 3.15):

$$\Delta\Phi = \Phi_a - \Phi_o = g\,\Delta h = g\,(h_a - h_o). \qquad (3.14a)$$

If, however, the effects of different rock pods are to be compared, it is useful to introduce two additional parameters, namely, the limit anomaly, $\Delta\Phi_l$ or Δh_l, and the relative anomaly, $\Delta\Phi_r$ or Δh_r.

The limit anomaly, $\Delta\Phi_l = g \cdot \Delta h_l$, is the maximum possible change in the fluid potential at each end of the lens. Its absolute value is:

$$|\Delta\Phi_l| = \frac{L}{2}\frac{\partial\Phi_o}{\partial x} = |g\,\Delta h_l| = g\frac{L}{2}\frac{\partial h_o}{\partial x}, \qquad (3.14b)$$

where x is the direction of the original potential gradient (Fig. 3.15).

The significance of the limit anomaly as a reference value becomes clear when it is realized that lens-induced anomalies of the fluid potential increase with lens permeability k' only to a limiting value of the permeability ratio $\varepsilon = k'/k$, where k is the permeability of the rock matrix surrounding the lens. In the case that $\varepsilon = k'/k \geq \varepsilon_l$, the lens' permeability may be considered infinitely large, $k \approx \infty$. Essentially, in such cases the lens no longer offers significant resistance to flow and the hydraulic gradient in it approaches zero. Those values of the fluid potential that occurred in the central portions of the lens in the undisturbed case migrate to the lens's extremities: the original values at the upstream end reduce, while at the downstream end they increase, respectively, to the limiting minimum and maximum values possible under the given set of parameters.

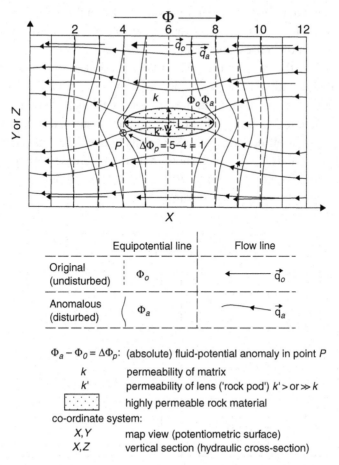

Fig. 3.15 Effect of relatively highly permeable rock lens on originally uniform fields of flow, $\mathbf{q}_0$, and fluid potential, Φ_0(Tóth and Rakhit, 1988, Fig.1, p. 363).

The *relative anomaly,* $\Delta\Phi_r$ or Δh_r, is defined as the ratio of the absolute anomaly at one end point of the lens, $\Delta\Phi_{L/2}$ or $\Delta h_{L/2}$, to the limit anomaly $\Delta\Phi_l$ or Δh_l:

$$\Delta\Phi_r = \frac{\Delta\Phi_{L/2}}{\Delta\Phi_l} = g\Delta h_r = g\frac{\Delta h_{L/2}}{\Delta h_l}. \qquad (3.14c)$$

The relative anomaly is a measure, therefore, of the actual effect of a permeable lens on the fluid-potential field, expressed in terms of the theoretically possible maximum distortion caused by a lens of the same geometry.

(i) Effect of permeability contrast and lens geometry. The effect of the permeability contrast ε and lens geometry L/W (length-to-width ratio) on the hydraulic head is analysed by calculating the perturbed potential fields h_a, absolute anomalies Δh,

and relative anomalies Δh_r, for different values of ε and L/W (Tóth and Rakhit, 1988). The example in Figure 3.16 demonstrates the effects on the hydraulic head field of changes in the lenses' sphericity, i.e. L/W ratio.

The length L of the longitudinal axis is kept parallel to the flow direction and constant while the width of the lens is increased. As the ratio L/W decreases from 10:1 to 5:4, both the areal extent and the intensity of the perturbation increase. The anomalies become increasingly elliptical in a direction normal to the original flow direction. The effect of increasing lens sphericity is even more strongly expressed by the relative anomaly, Δh_r, which shows a marked increase with increasing lens width W for any given conductivity ratio ε, from $\varepsilon = 1$ (homogeneous rock), to the limiting value of $\varepsilon = \varepsilon_l \approx 1000$ (Fig. 3.17).

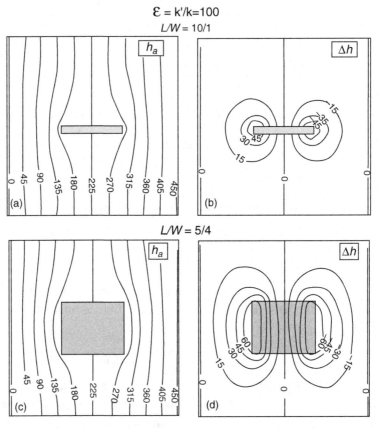

Fig. 3.16 (a, c) Calculated distributions of hydraulic head, h_a and (b, d) corresponding hydraulic-head anomalies, Δh produced in a homogeneous field of hydraulic gradient $\partial h/\partial x = 0.02$ by permeable lenses of permeability contrast $\varepsilon = 100$ and different length-to-width ratios, L/W. Contour values are in arbitrary relative units (after Tóth and Rakhit, 1988, Fig. 3, p. 365).

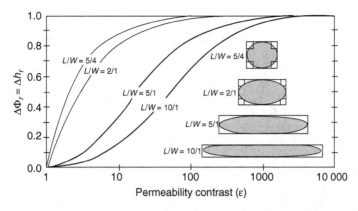

Fig. 3.17 Relative anomaly of the fluid potential, $\Delta\Phi_r$ (or hydraulic head Δh_r) as a function of length-to-with ratios, L/W and permeability contrast, ε, of ellipsoidal lenses placed in a uniform field of hydraulic gradient $\partial\Phi/\partial x = \partial h/\partial x = 0.02$ (analytical solution; after Tóth and Rakhit, 1988, Fig. 4, p. 365).

The effect of changes also in the permeability contrast ε is clearly seen in Figure 3.17: an increase in $\varepsilon = k'/k$ up to a value of approximately $\varepsilon = 1000$ results in increases in the relative anomalies. Beyond this value, however, the relative anomaly closely approximates the theoretically possible maximum. Further increases in the lens' permeability (and thus in ε) cannot produce significant additional distortions in the hydraulic head field. It should be noted that with increasing sphericity the permeability contrast needed to reach the maximum anomaly decreases. Conversely, probably no practically significant increases are induced in possible maximum anomalies by permeability contrasts beyond $\varepsilon = 1000$, if the ratio $L/W \geq 10$. As mentioned earlier a, similar conclusion was also reached by Freeze and Witherspoon (1967) in connection with the basin-scale effects of highly permeable strata [p. 31, Sec. 3.2.1, a(ii)].

(ii) Effect of lens orientation. The orientation, α, of a rock lens relative to the undisturbed direction of the hydraulic-head gradient $(\mathrm{grad}h_0)_x$ has characteristic effects on the distribution patterns of both the hydraulic head h_a and anomaly Δh. Figure 3.18(a–f) illustrate such effects for three different positions of a rock lens, namely, for $\alpha = 0°, 45°$ and $90°$. As a result of the lens's rotation from a parallel to a perpendicular position (Figs. 3.18 a–f, respectively) relative to the original flow direction $(\partial h/\partial x)_0 = (\mathrm{grad}h_0)_x$ the areal extent and the intensity of the perturbation decrease, as seen from the patterns and values of both the equipotentials and the anomalies (Figs. 3.18a,c,e and 3.18b,d,f, respectively). In an oblique position (e.g., Figs. 3.18c,d), which is probably the most common situation in reality, the major perturbations of the potential field are displaced laterally and in opposite directions

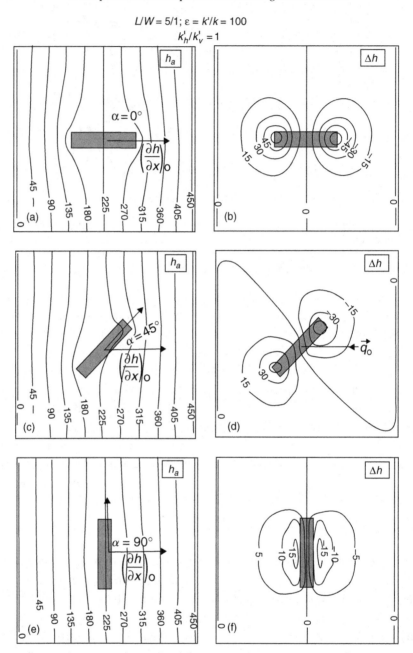

Fig. 3.18 Calculated distributions of hydraulic head $h_a(a, c, e)$ and corresponding hydraulic-head anomalies $\Delta h(b, d, f)$ produced in a homogeneous field of hydraulic gradient $\partial h\,\partial x = 0.02$ by a permeable lens of length-to-width ratio $L/W = 5/1$, permeability contrast $\varepsilon = 100$, and with angles α between its long axis and the direction of original gradient $\partial h/\partial x$ of $0°45°$and $90°$. Contour values are in arbitrary relative units (after Tóth and Rakhit, 1988, Fig. 5, p. 366).

with respect to the original gradient $(\mathrm{grad}h_0)_x$ (Figs. 3.18c,d). Clearly, the more closely the long axes, for instance, of a sand bar or channel sand are aligned with the undisturbed direction of the formation-fluid flow, the greater is the accuracy by which their positions and dimensions may be deduced from observed potentiometric perturbations.

(iii) Effect of multiple lenses. When two or more rock pods are placed closer together than the combined radii of their potentiometric influence, the perturbations coalesce and the resulting total anomaly is different from one that would be caused by a single lens. From numerous possible situations (Tóth and Rakhit, 1988), one example is presented in Figure 3.19(a,b): two identical lenses in parallel positions and with partial overlap along their lengths. The pair of positive and negative anomalies associated with the overlapping, i.e. proximate, ends of the lenses might be (mis-) interpreted to indicate a single lens. The long axis of such a lens would be nearly normal to those of the real lenses. This interpretation would, however, also require a flow field that is oriented approximately parallel to an imaginary line joining the foci of the positive and negative anomalies. It could, therefore, be readily invalidated by a potentiometric map, which indicates the actual flow direction.

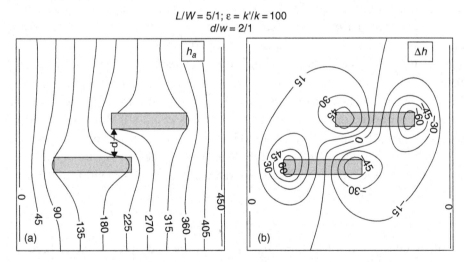

Fig. 3.19 (a) Calculated distributions of hydraulic head, h_a and (b) corresponding hydraulic-head anomaly, Δh, produced in a homogeneous field of hydraulic gradient, $\partial h/\partial x = 0.02$, by two identical permeable lenses in parallel position, partial overlap; length-to-width ratio $L/W = 5/1$, permeability contrast $\varepsilon = 100$, longitudinal axes parallel to undisturbed gradient directions, lens separation, d, two width units. Contour values are in arbitrary relative units (after Tóth and Rakhit, 1988, Fig. 9, p. 369).

In summary, the lens-scale effects of highly permeable rock bodies on formation-water flow may be characterized by the following intrinsic properties (Tóth and Rakhit, 1988):

(i) the potentiometric anomaly, $\Delta\Phi$ or Δh, is negative at the upstream end and positive at the downstream end of a highly permeable lens;

(ii) the absolute value, $|\Delta\Phi|$ or $|\Delta h|$, of the anomaly increases with increases in any of the following parameters: permeability contrast $\varepsilon = k'/k$, length L, width W, or length-to-width ratio L/W of the lens;

(iii) the absolute value $|\Delta\Phi|$ or $|\Delta h|$ of the anomaly cannot exceed $\left|\frac{L}{2}\frac{\partial\Phi_o}{\partial x}\right|$ or $\left|\frac{L}{2}\frac{\partial h_o}{\partial x}\right|$.

3.2.3 Effects of faults

A *fault* is a fracture or zone of fractures caused by movement of rock blocks relative to one another on both sides of its predominantly planar and parallel surfaces. Movement along a fault's opposite surfaces may stop and resume several times during the geologic history of a region. In many instances, faults represent discontinuities in permeability and may thus affect the flow of formation waters.

The effects of faults on formation-fluid flow can vary greatly, from being insignificant to determinant. They depend on numerous factors, the most important of which include: permeability and thickness of the fault's body; permeability of the strata juxtaposed across the fault; fluid-potential values at the fault's termination and, in particular, whether or not the fault outcrops at the land surface; the hydraulic regime (recharge, midline, discharge) of the fault's outcrop area, if any; the dip angle of the fault; and the angle of incidence between flow lines and the fault's surface.

A comprehensive discussion of the possible types and roles of faults in basinal hydraulics is beyond the scope of this book. [Jones *et al.*, (1998) contains several relevant papers; Underschultz *et al.*, (2005) present some general concepts and field examples from Western Canada and the North West Shelf of Australia]. Instead, two basic hydraulic types of faults are briefly and qualitatively mentioned, namely: barrier faults and conduit faults.

3.2.3.1 Barrier faults

A fault is considered a *hydraulic barrier* when its body (i.e. the rock between its bounding walls: fault breccia, fault gouge), has a contrastingly lower permeability than the formations juxtaposed across it. As an end-member of such situations a barrier fault functions and may be modelled as an impermeable boundary. Impermeability of a fault, however, is only a theoretical limiting case. It seldom, if ever, exists in nature. The assumption of absolute impermeability of faults may lead to erroneous conceptualizations of basinal flow patterns and their evolutionary history, thus to false conclusions regarding the flow's expected consequences. The barrier effect of faults on basinal flow can be reduced by possible temporal changes in a fault's permeability. Such changes may be brought about by mechanical and/or chemical alterations of the fault's body or its surface, by dissolution and removal of gouge material by formation fluids, or by changes in relative positions of the aquifers and aquitards on both sides of the fault plane.

Two basic effects on groundwater flow of single, idealized barrier faults in a unit basin can be identified and are illustrated in Figures 3.20(a–d), namely: (i) compartmentalization, and (ii) hydraulic sheltering.

(i) Compartmentalization. Barrier faults divide the flow domain into sections in which the flow is controlled by the sections' own boundary conditions. Vertical and high-angle, or outcropping, barrier faults result in the development of new basins of reduced horizontal dimensions and positioned side-by-side (Figs. 3.20a,c,d).

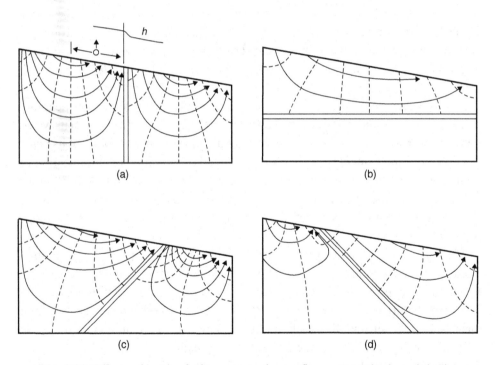

Fig. 3.20 Effects of barrier faults on groundwater flow patterns in the unit basin: (a) vertical fault; (b) horizontal fault; (c) fault dipping opposite to direction of regional flow; (d) fault dipping in direction of regional flow.

Hydraulic heads drop step-wise across the barriers in the direction of the flow, as is indicated by a crowding of the equipotential lines in a band along the fault on the potentiometric map. New areas of discharge develop, possibly marked by flowing artesian conditions, adjacent to the upstream sides of new areas of recharge downslope from the fault. Horizontal and low-angle faults, i.e. non-outcropping, barrier faults, compartmentalize the flow domain in a vertical sense: the depth of the upper zone of strong flow is reduced, and the water becomes stagnant in the isolated basal zone (Fig. 3.20b).

(ii) Hydraulic sheltering. Each new flow cell generated by barrier faults is accompanied by additional quasi-stagnant zones. Consequently, the relative volumes of water that are 'sheltered' from active circulation, as compared with undivided basins, are thereby increased. The sheltering effect is particularly strong beneath the foot wall of a barrier fault. The effect seems to increase with a decrease in the fault's dip and it reaches maximum beneath a horizontal fault (Fig. 3.20b).

3.2.3.2 Conduit faults

A fault is considered a hydraulic conduit if it enhances fluid flow through it. Two major types of conduit faults can be distinguished based on the relative direction of flow path, namely: a *conductive conduit* fault and a *barrier conduit* fault.

A fault is *conductive conduit* if its body has a contrastingly higher permeability than the formations around it and the bulk of flow is parallel to and between its walls. It is, however, a *barrier conduit* if, notwithstanding the low permeability of its body, the relative displacement of its walls has resulted in hydraulic communication of permeable formations juxtaposed on its opposite sides and the bulk of flow is thus normal to its walls. Flow patterns in and around the barrier conductive faults are so diverse and complex as to defy any attempt at generalization within the scope of the present work.

The basic effect of a conductive conduit fault can, however, be briefly summarized: it is the collection and concentration of flow into the fault zone (Figs. 3.21a–d). Essentially, the effect is the same as that of an aquifer embedded in a relatively low-permeability matrix. It depends on the permeability contrast between the fault material and the surrounding rock, on the thickness of the fault, and on the angle of incidence between the flow lines and the fault plane.

The effect of a conductive conduit fault on the flow field is virtually nil if the angle of incidence is $\alpha \approx 90°$: the flow lines cross the fault un-refracted (Fig. 3.21a). If the angle is different from 90°, part of the water is deflected into and along the fault:

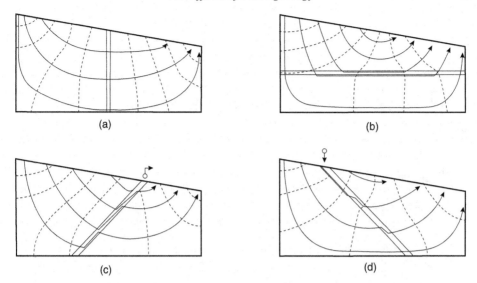

Fig. 3.21 Effects of conduit faults on groundwater flow patterns in the unit basin: (a) vertical fault; (b) horizontal fault; (c) fault dipping opposite to direction of regional flow; (d) fault dipping in direction of regional flow.

the flux in the fault increases (Figs. 3.21b–d). If the fault outcrops then, depending on the dip direction relative to the flow, part of the water may be 'shunted' to the surface where it may appear as fault springs, whilst flow intensity is reduced on the fault's footwall side (Fig. 3.21c). Alternatively (Fig. 3.21d), additional recharge may be induced through the outcrops resulting in increased flow intensity through the basin on the downstream side of the fault. A slight reduction in flow intensity, accompanied by an expansion of the quasi-stagnant zone, occurs beneath the foot-walls of the conductive faults too. The extent of the reduction is less, however, than it is in the case of the barrier faults: the reduction is caused now by water being shunted away from hydraulically continuous parts of the basin, rather than those parts being hydraulically isolated. The difference becomes important in terms of consequences should boundary conditions change during the basin's evolutionary history.

3.2.4 Effects of anisotropy

The rock framework's permeability is generally anisotropic, i.e. its permeability (hydraulic conductivity) has different magnitudes in the three coordinate directions, $k_x \neq k_y \neq k_z$. Anisotropy results in increased flux in the direction of higher k value. A basin-wide effect of this differentiation of flow-intensity is demonstrated by Freeze and Witherspoon (1967). Using their homogeneous-isotropic

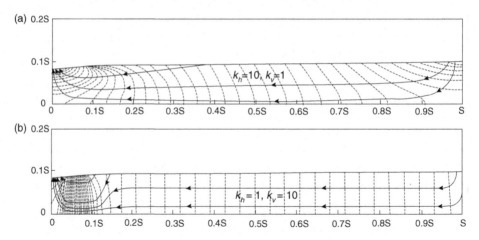

Fig. 3.22 Effect of anisotropy on regional groundwater flow patterns, linear basin surface, broad central valley: (a) $k_h/k_v = 10$; (b) $k_h/k_v = 1/10$ (after Freeze and Witherspoon, 1967, Fig. 6A,B, p. 631).

model (Fig. 3.7a) as the reference case, Figure 3.22(a) shows that an increase in the horizontal permeability, k_h, relative to the vertical permeability, k_v, to a ratio of $k_h/k_v = 10$ results in an increased area of recharge and in less arcuate, horizontally stretched, flow trajectories. A relative increase in vertical permeability, k_v, to a ratio of $k_h/k_v = 1/10$ (Fig. 3.22b), on the other hand, accentuates the vertical flow components, reducing thereby the areas of descending flow and extending the lateral-flow segment of the flow field.

An impression of the local-scale effects of anisotropy is obtained by comparing the potentiometric perturbations and anomalies caused by a homogeneous-isotropic lens, $k_h'/k_v' = 1$ (Figs. 3.18a,b), with those of two similar rock bodies, but with horizontal-to-vertical permeability ratios k_h'/k_v' of 2 and 10 (Figs. 3.23a,b and c,d, respectively (Tóth and Rakhit, 1988)). It appears that an increase in horizontal permeability reduces the vertical extent of the potential field's perturbations and, at the same time, extends it laterally (compare Figs. 3.18a and 3.23a,c). The effect is emphasized by the anomaly contours, as their shapes change from an ellipse with long axis normal to flow in the isotropic case (Fig. 3.18b), to ellipses increasingly elongated in the direction parallel to flow, as the degree of anisotropy increases (Figs. 3.23b,d).

One practical significance of these observations is that in beds or strata of common anisotropy type, for instance, in argillaceous or siliciclastic rocks, where $k_h >$ or $\gg k_v$, lens-induced potentiometric anomalies tend to be confined to within the unit's boundaries. The possibility of interference between lens-induced anomalies in the vertical direction, i.e. cross-formationally, is thus reduced by anisotropy.

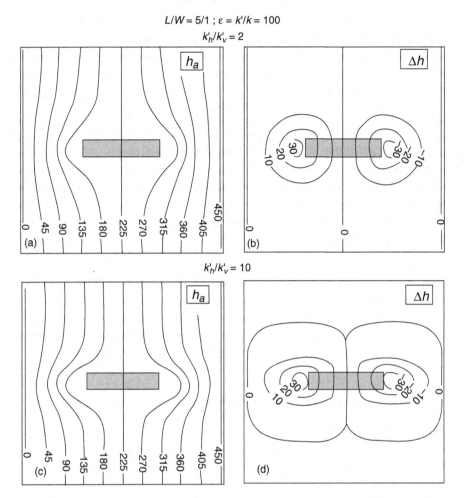

Fig. 3.23 (a, c) Calculated distributions of hydraulic head, h_a and (b, d) hydraulic-head anomalies, Δh, produced in a fluid-potential field of hydraulic gradient, $\partial h / \partial x = 0.02$, by a permeable lens. The lens's properties are: length-to-width ratio $L/W = 5/1$, permeability contrast $\varepsilon = 100$, longitudinal axis parallel to the undisturbed gradient, degrees of anisotropy, k_h/k_v, 2 and 10. Contour values are in arbitrary relative units (after Tóth and Rakhit, 1988, Fig. 6, p. 367).

3.3 Effects of temporal changes in the water table: transient pore pressures and flow systems

Temporal modifications in the shape of the water table at different scales are common in nature. Such changes may be due to a number of different geologic factors such as erosion, uplift, rifting or crustal warping, but also to changes in seasonal precipitation, weather pattern or climate. Their time rates, durations and magnitudes vary over widely different ranges.

In turn, a change in the water table disturbs the current equilibrium between the boundary conditions and the flow field in the domain. It induces changes in pressure, hydraulic head, and flow, striving to reestablish steady-state patterns adjusted to the new boundary conditions. During the time of adjustment the various dynamic parameters and their patterns of distribution are transient, i.e. in disequilibrium. Depending on the time rate of change in the boundary conditions relative to the rate of adjustment inside the flow domain and the duration of the process, different degrees of disequilibrium develop between the boundaries and the pressure and flow fields. Symptomatic of such disequilibria are abnormal pore–fluid pressures.

In basinal hydrogeology, the most important consequences of temporal changes in the water table are: (i) transient pore pressures; (ii) changes in flow patterns; (iii) time-dependent water-flow rates; (iv) anomalies in flow-generated or flow-sensitive processes and phenomena. Such processes and phenomena may include, e.g., chemical composition of groundwater, subsurface temperature fields, accumulation and dissipation of matter, and so on. Whether or not an anomaly exists in reality is inferred from 'discrepancies' between observable field conditions and theoretically expected consequences of the flow systems. The basis of analysis, interpretation, and modelling of transient basinal flow conditions, and their expected consequences, is the diffusion equation (Eq. 1.1), presented earlier.

3.3.1 Time lag and time scales in pore-pressure adjustment

According to the diffusion equation,

$$\frac{\partial^2 h}{\partial x^2} + \frac{\partial^2 h}{\partial y^2} + \frac{\partial^2 h}{\partial z^2} = \nabla^2 h = \text{div grad } h = \frac{S_0}{K}\frac{\partial h}{\partial t}, \tag{1.1}$$

if the hydraulic head changes with respect to time in an elementary volume, i.e. when $\partial h/\partial t \neq 0$, the change results in a finite difference between the flow rates into and out of the volume element. The volume functions as a source or a sink for the fluid involved. The difference in the flow rates is controlled by the terms $\partial^2 h/\partial x^2$, etc., and by the diffusion coefficient $D = K/S_0$.

A change in the inflow–outflow balance in one part of the flow region does, of course, disturb the flow direction and flow intensity in other, hydraulically connected, points. The effect is manifested by changes in the hydraulic heads: a head (or pressure) disturbance propagates through the voids of the rock framework to other points by diffusion. The rate of this pressure dissipation is a function of the coefficient of diffusivity, D. In this coefficient, the hydraulic conductivity K represents the ease with which a fluid flows through the rock. The rate of pressure and fluid-mass transfer is, therefore, directly proportional to K. The specific storage, S_0, on the other hand, is a measure of the fluid volume that a rock/fluid system can absorb or release by a unit change in hydraulic head due to elastic changes in the volume of water and pore space. Consequently, the rate of pressure and fluid-mass transfer is inversely proportional to S_0.

Pressure changes can thus be expected to propagate more rapidly in highly permeable (high K) and slightly deformable (low S_0) rigid rocks, such as indurated porous sandstones, than in poorly permeable and highly deformable rocks as, for instance, clays or incompletely compacted shales. In all cases, however, because of the finite value of the coefficient of diffusivity, a time lag develops between an initial, inducing change in pore pressure (the cause) at one point, and the response of induced changes in pore pressure (the effect) at all other points of the flow region. The time lag increases with increasing distance between the points of origin and observation, with increasing specific storage S_0, and with decreasing hydraulic conductivity K.

Whether or not the question of lag time needs to be included in the considerations of a given problem depends on the time scale of the expectable pore-pressure response to a perturbation, relative to the time scale of the problem itself. The time scales of both the propagation of pore-pressures in nature and the related natural processes of human interest range over many orders of magnitude. Consequently, in certain problems the two time scales do not overlap. Ignoring the transient nature of the problem is justified in these cases. On the other hand, ignoring the fact that a finite length of time is required for pore-pressure disturbances to traverse certain distances in the rock framework can lead in many instances to incomplete or erroneous interpretations, conclusions, and practical decisions in flow-related questions. The two examples presented below are intended to illustrate a number of points related to transient pore-pressure conditions. These include: the often ignored, even denied, fact that pore-pressure disturbances do propagate through rocks of very low permeability; the wide range of time scales possibly relevant to practical problems such as subsurface waste disposal or petroleum hydrogeology; the fact that the time required for pore-pressure changes initiated at one point to cause observable changes in other points may exceed the duration of observation; potential errors that can be made by ignoring transient conditions; and the fact that transient pore pressures exist on the geological time scales, in general, and that changes in topographic configuration can induce them, in particular.

The first example illustrates the phenomenon of time lag in pore-pressure responses on the human time scale. As part of the evaluation of a potential gas-storage reservoir, Witherspoon and Neuman (1967) determined the permeability of a 16-foot (≈ 4.9 m) thick shale by observing the response of water levels in the Galesville Sandstone aquifer above the shale, to pumping from the potential reservoir, below the shale (Fig. 3.24). A permeability of $k = 0.710^{-4}$ md

$(K \approx 0.7 \times 10^{-12}$ m/s) was obtained from the pumping test, in good agreement with core-analysis results yielding $k = 1.8 \times 10^{-4}$ md $(K \approx 1.8 \times 10^{-12}$ m/s). However, had pumping and observations stopped before 30 days (in practice, few pumping tests are conducted for more than a few days) the 16 foot $(\approx 4.9$ m) thick caprock would have appeared impermeable and the two sandstones above and below it hydraulically unconnected. In the authors' own words (Witherspoon and Neuman, 1967, p. 954): 'it was apparent that there is no evidence of drawdown until after about 30 days (of pumping). An unmistakable downward trend began after 40 days and continued for 20 days after the water withdrawal stopped. About 45 days after the pumping ceased, the fluid levels began to recover.' (Fig. 3.24).

In the second example, delays in cross-formational responses to pressure changes on the geological time scale are demonstrated. Such lag times are beyond the possibility of direct observation on the human time scale and estimates can only

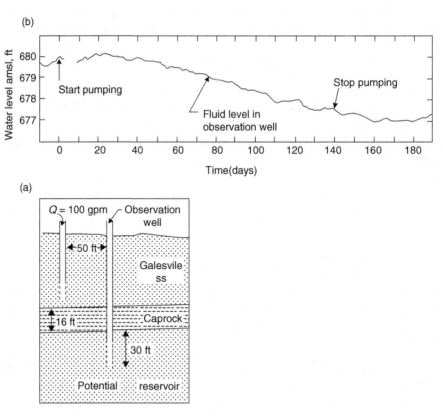

Fig. 3.24 Observed lag of water-level changes caused by pumping across thin caprock (modified from Witherspoon and Neuman, 1967, Figs. 1 and 7, p. 950 and 953, respectively; conversion of values shown on figure: 50 ft $\approx$ 15.3 m, 16 ft $\approx$ 5 m, 30 ft $\approx$ 9 m, 680 ft $\approx$ 208 m, 677 ft $\approx$ 206 m, 100 gpm–US gallons per minute– $\approx$ 6.3 l/s $\approx$ 380 l/min).

be based on theory and calculations. In this case, times were calculated for pore-pressure adjustments at the base of an approximately 1000 m thick sedimentary rock column including slightly permeable strata to changes in the water table, for a realistic geological situation in northern Alberta, Canada (Figs. 3.25a,b; Tóth, 1978; Tóth and Millar, 1983).

The coefficient of diffusivity $D = K/S_0$ was varied in Equation 1.1 by varying the hydraulic conductivity, K, while keeping the specific storage constant at $S_0 = 10^{-6}/cm$ for all strata. The required adjustment times t_a were expressed as a function of the relative adjustment of hydraulic head, R_{an}. It is defined as the ratio of the induced head-change actually accomplished at depth, to the inducing total change at the land surface:

$$R_{an} = \frac{\Delta h_{an}}{\Delta h_T}. \tag{3.15}$$

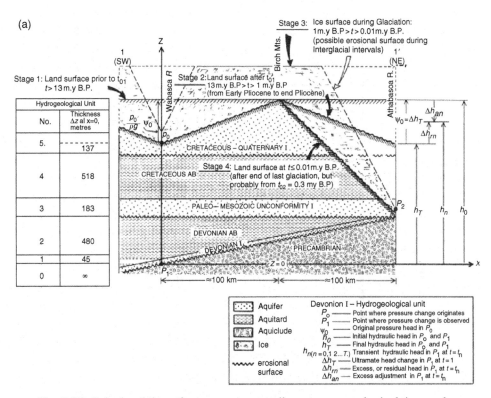

Fig. 3.25 Calculated times for pore-pressure adjustment on geological time scale, Red Earth region, northern Alberta, Canada: (a) idealized hydrogeological cross-section and evolutionary stages used in calculating hydraulic-head changes at P_1 in Devonian I Aquifer; (b) hydraulic-head changes at the base P_1 of a sedimentary column generated by an instantaneous head change at the land surface P_0, as a function of the hydraulic parameters of hydrogeological units 2 and 4 (after Tóth and Millar, 1983, Figs. 3 and 4, p. 1587 and 1590, respectively).

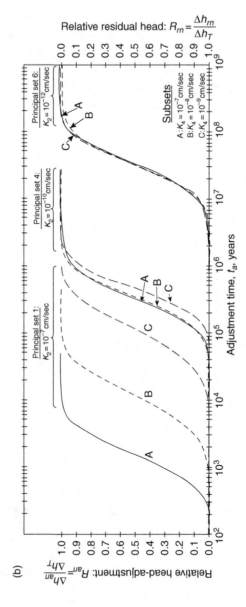

Fig. 3.25 (cont.)

In Equation (3.15) Δh_{an} is the actual change in head at the point of observation at time t_n, and Δh_T is the total change in head at the water table.

Figure 3.25b shows that as the hydraulic conductivity K_2 of the lower aquitard decreases through the values $K_2 = 10^{-7}$, 10^{-10} and 10^{-12}cm/s, (principal sets 1, 4, 6) the onset time t_{a0} (the time required for the first noticeable change in head at depth) increases through the values of 2×10^2, 4×10^4 and 4×10^6 years, respectively, while the conductivity of the upper aquitard is kept at a constant value of $K_4 = 10^{-7}$cm/s (curves 1A, 4A, 6A). The respective times for complete adjustment t_{a100} are approximately 1×10^4, 2×10^6 and 2×10^8 years. The first indication, or 'breakthrough', and the transmittal of 50 and 100 per cent of the hydraulic head change are delayed by 7×10^4, 4.7×10^5 and 2×10^6 years, respectively, by a combination of two thick strata with the slight, but in nature common, values of $K_2 = 10^{-10}$ and $K_4 = 10^{-9}$cm/s (curve 4C). Many geologic phenomena and processes may be affected by changes in subsurface flow lasting for time spans mentioned above, such as: regional patterns of temperatures and ionic and isotopic constituents of formation waters; migration and accumulation of hydrocarbons and metals; diagenetic changes of silicate and carbonate minerals.

3.3.2 Effect on basinal flow patterns

Delays in pore-pressure responses to changes in boundary conditions may affect the basinal fields of pressure, hydraulic head and, thus, flow and its natural con-sequences. Awareness of these effects is useful and can be important in applied hydrogeology. One potentially puzzling effect is a possible discrepancy, i.e. dis-equilibrium, between observed fields of dynamic parameters and their expected patterns derived theoretically from the current topography and assumed steady-state conditions. A corollary to this effect is the possible inconsistency between the actual flow patterns and various flow-generated or flow-sensitive natural pro-cesses and phenomena. While abnormal pore pressures are known possibly to result from time lag in pressure adjustment to a relatively rapidly changing topographic relief, they are also frequently misinterpreted as steady-state anomalies caused by 'impermeabilty' of strata or faults, or both.

Figures 3.26(a–c) illustrate schematically the possible effects that an inversion of the topographic relief can have on basinal water flow in the presence of a stratum of low diffusivity (Tóth, 1984; Tóth and Corbet, 1986). On a basinal scale, the water table is assumed to coincide with the land surface. Modification to the topographic relief begins at time t_1 (Fig. 3.26a). At this time the land surface is S_1. Surface S_1 has already been in existence for a sufficiently long time prior to t_1 that the contem-poraneous flow systems, q_1, are perfectly adjusted to it, thus they are in steady-state equilibrium with the topography. At time t_1 the land surface begins to change and

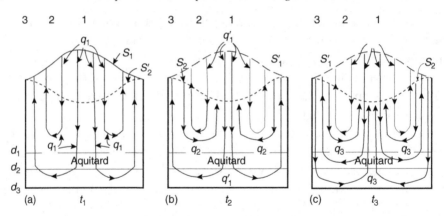

Fig. 3.26 Conceptual illustration of delayed adjustment of flow systems to temporal changes in water-table relief: (a) original, t_1, steady-state flow-systems, monochronous flow field; (b) transient, t_2, flow systems, heterochronous flow field; (c) final, t_3, steady-state flow systems, monochronous flow field. Current surface: solid line; future and paleo surface: dashed line (Tóth, 1995, Fig. 8, p. 12; modified from Tóth, 1988, Fig. 22, p. 498).

by time t_2 the former ridge S_1 becomes the new valley of S_2 (Fig. 3.26b). If the rate of topographic change is greater than the possible rate of cross-formational pressure-diffusion from the zone below the aquitard, a flow pattern develops that is incongruous with the relief. The now-modern flow systems q_2 in the high diffusivity upper parts of the basin follow the changes in the topography and remain adjusted to the boundary conditions. As a result, the original systems q_1 are eliminated in this upper zone: they are rendered to the status of paleo-, i.e. extinct, flow systems q'_1. Due to the low diffusivity of the aquitard, however, remanent differences in fluid potentials (and pore pressures) keep driving the fluids centrifugally, thus maintaining the relict, gradually decaying and therefore transient, q'_1 systems for a while in the lower portions of the flow region.

Finally, at time t_3 (Fig. 3.26c), after a sufficiently long time has elapsed for the relict potential differences in the aquitard to dissipate, the modern, continually renewed systems, q_3, prevail throughout the entire depth of the basin. By now, all previous flow systems become 'paleo', i.e. extinct. At this time, all flow systems in the basin are perfectly adjusted to the current land surface S_2 and are, therefore, in steady state. They are, however, completely opposite in direction to what was their original orientation at time t_1. In between, at time t_2 (Fig. 3.26b), the incongruous situation existed that shallow and deep systems were oppositely directed, with the deep systems flowing centrifugally toward topographically high regions. In view of the adjustment times, discussed in the previous section, such a puzzling situation appears to be entirely possible in a fluid regime which is hydraulically continuous and driven solely by gravity. This observation is not meant, however, to propose that gravity is the sole driving force in *all* such situations!

One possible, but *erroneous*, explanation for the basinal flow pattern at time t_2 could envisage a hermetically sealed aquifer in which unbalanced compaction, thermal, or osmotic forces cause over-pressuring and drive centrifugal flow. The sealing aquitard would be overlain by a basin with a gravitational flow regime completely separated from the deeper system. In reality, however, the entire basin is one hydraulically continuous unit, which has experienced a change in the relief of the water table and is now undergoing hydraulic adjustment, controlled by the diffusivity of a major aquitard.

Indeed, the diffusivity even of a major aquitard is only one of the important controls on the adjustment of pore pressures. In a sensitivity analysis of a two-dimensional version of the previously discussed example from northern Alberta (Fig. 3.25a,b; Tóth and Millar, 1983), England and Freeze (1988) examined the effect of variations in several factors on the rate of hydraulic head decay, namely: hydraulic conductivity, anisotropy, specific storage, water-table configuration and basin size. They found that adjustment times are, in general, relatively highly sensitive to changes in: hydraulic conductivities of the poorly permeable strata, vertical components of K if the rock is anisotropic, specific storage of the aquifers, and thickness of the basin, i.e. the vertical dimensions of the strata. On the other hand, adjustment times show less sensitivity to changes in: hydraulic conductivity of the highly permeable strata, the magnitude of the horizontal component of K in case of anisotropy, specific storage of poorly permeable rocks, water-table configuration and the length of the basin.

In a study aimed at evaluating the suitability of two sites possibly considered for high-level nuclear fuel-waste repositories, Senger *et al.* (1987) examined the effects that topographic modifications due to tectonic and geomorphic processes may have had on the subsurface flow systems since Miocene times in the Palo Duro Basin of West Texas, USA. The analysis was conducted by transient numerical modeling along an approximately 600 km long west–east section crossing the High Plains from New Mexico to the Eastern Caprock Escarpment. It simulated four major stages in the basin's evolutionary history: (i) uplift and tilting; (ii) deposition of the surficial Ogallala aquifer; (iii) erosion of the Pecos River valley; (iv) the westward retreat of the Eastern Caprock Escarpment. In the series of successive simulations, the final distribution of hydraulic heads for each stage was used as the initial condition for the subsequent step.

The basin is divided into three major hydrogeologic units: the Deep Basin Brine aquifer at the base; the Ogallala and Dockum aquifers at the top; and, between and separating the two units, a thick aquitard made up of Permian evaporites. During all four stages of the topographic modifications, significant adjustments in hydraulic heads are observed in the Deep Basin Brine aquifer 1 Ma after the change in topography, and near or fully steady-state conditions develop within a maximum of 9 Ma. In particular, in simulating the effect of the erosion of the Pecos River valley, hydraulic heads were gradually lowered during a period of 4 Ma. By 5 Ma after the start of the erosion, i.e. 1 Ma after it has ended, virtually steady-state conditions prevailed, with only about 1.7 m maximum head change occurring over the last 1 Ma.

The approximate distributions of steady-state hydraulic heads and stream lines 5 Ma after the onset of the erosion of the Pecos River valley, and the effect of the erosion in terms of hydraulic-head differences before and thereafter, are shown in Figures 3.27(a,b) respectively (Senger *et al.*, 1987, Figs. 12 and 25). Figure 3.27(a) shows that, in spite of the deeply penetrating effect of the Pecos River, a deep regional flow system does pass beneath its valley. The flow rate of the deep system is, however, only a fraction of that of the shallow systems (note values of the streamline contour intervals). Figure. 3.27(b) illustrates the magnitude and areal extent that major changes in topography can have on the hydraulic head distribution, thus

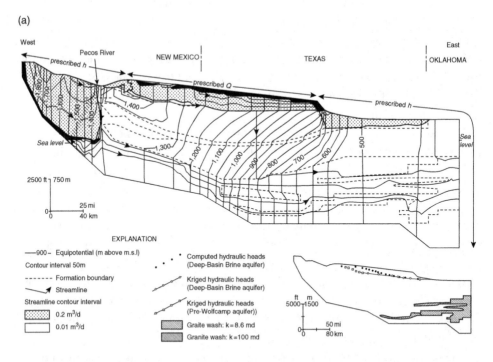

(a)

Fig. 3.27 Simulation of transient flow fields on geologic time-scale for the Palo Duro Basin, West Texas, USA: (a) subhydrostatic conditions in the Deep-Basin Brine aquifer east of the Eastern Caprock Escarpment; (b) difference in hydraulic heads before and after Pecos River valley erosion (modified from Senger *et al.*, 1987, Figs. 12 and 25, respectively)

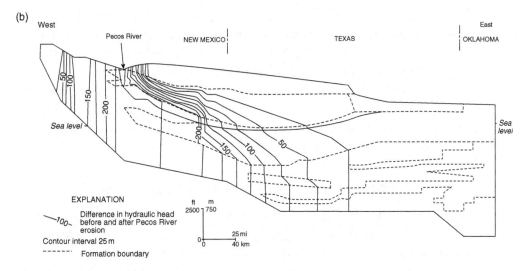

(b)

Fig. 3.27 (cont.)

also on the direction and intensity of flow, even beneath very slightly permeable $(2.8 \times 10^{-4}$ md $\approx 2.8 \times 10^{-12}$ m/s$)$ and thick (several hundred metres) evaporite aquitards, within less than 5 million years.

3.4 Hydraulic continuity: principle and concept

The theories, their practical applications, and the interpretations of field observations discussed so far were based on the assumption that the rock framework is hydraulically continuous. The assumption was justified tacitly by recognizing that: (i) the existence of large-scale groundwater flow systems and regionally extensive patterns of hydrogeologic phenomena are confirmed by observations; (ii) the effects of naturally and artificially imposed stresses on the groundwater body propagate over great distances and depths; (iii) the rate of sustainable water production depends ultimately on some weighted average permeability of all the rocks included in a basin-wide 'representative elementary volume' ('REV'), or even on available recharge, rather than on the pumped aquifer's transmissivity T and storage coefficient S; (iv) industrial, agricultural, and municipal contaminants of known sites of origin may appear at unexpected distances, locations and depths. Indeed, one even might argue that the primary, although not necessarily obvious, justification for considering large sedimentary basins as natural units for hydrogeological analysis is the fact that their different and often distant regions and depths are linked through a hydraulically communicating network of voids.

This fact, however, has not been generally recognized or accepted. For instance, one of the most eminent petroleum geologists of his day explained petroleum migration and entrapment, and several associated pressure, temperature, and mineralogical conditions by invoking hermetically sealed three-dimensional rock compartments (Hunt, 1990). While Hunt's ideas have gained some acceptance, the processes and phenomena discussed by him can also be understood on the basis of a hydraulically-continuous rock framework (Tóth *et al.*, 1991). The concept, with some supporting arguments, of regional hydraulic continuity as well as its natural consequences and effects are, therefore, reviewed below in the context of gravity-driven groundwater flow systems.

3.4.1 The concept of regional hydraulic continuity

Hydraulic continuity is a phenomenological property of the rock framework. It can be characterized quantitatively as the ratio of an induced change in hydraulic head (or pore pressure) to an inducing change in head (or pressure). Because pore-pressure changes propagate through the rocks at finite velocities (as controlled by the hydraulic diffusivity $D = K/S_0 =$ hydraulic conductivity/specific storage; Eq. (1.1)), whether or not hydraulic continuity can be perceived depends on the distance, and the travel-time of a pressure disturbance, between the points of origin and observation. Consequently, its demonstrability is a function of the scales of space and time at which a given problem is treated. In other words, it is a relative property.

The concept of hydraulic continuity is particularly useful in characterizing the hydraulic behaviour of heterogeneous rock masses on specified spatial and temporal scales, in general, and on the regional and geological segments of the space- and time-scale spectra, in particular. 'Thus, a subsurface rock body is considered hydraulically continuous on a given time scale if a change in hydraulic head at any of its points causes a head change at any other point, within a time interval that is measurable on the specified time scale' (Tóth, 1995, p. 6).

Regional hydraulic continuity is not a self-evident or easily verifiable property of the rock framework. Large contrasts in permeabilities of contiguous rock bodies may make the less permeable ones appear impervious from conventional types of observations. Pore-pressure responses at various points in the flow region to a pressure change elsewhere may take longer than the time span of observation, thus rendering the rock body, or parts of it, to appear impermeable. Or, where major contrasts in water chemistry, temperature or other flow-sensitive fluid properties coincide with low-permeability rock boundaries, an impression of hydraulic discontinuity is created or reinforced. It is, therefore, not entirely surprising that the century-long debate about hydraulic continuity, i.e. about the existence of absolutely isolated portions of the rock framework, has not yet been settled. The evolution of the concept and the principal arguments in support of hydraulic continuity are best reviewed by means of a brief historical retrospective.

3.4.1.1 Evolution of the concept

Current understanding of hydraulic continuity has developed independently from two different, indeed, opposite types of groundwater investigations, namely: aquifer and well hydraulics (local pumping tests) and basin hydraulics (regional water resources).

(i) Hydraulic continuity inferred from aquifer hydraulics. Chamberlain (1885), noted in his discussion of 'The Requisite and Qualifying Conditions of Artesian Wells', that 'No stratum is entirely impervious.' Nevertheless, authors of all early calculations of flow to wells assumed either a free water table or ideally confined conditions (e.g., Thiem, 1906; Theis, 1935, respectively). Hydraulic communication across confining strata was not formally taken into account until Hantush and Jacob (1954, 1955) included a 'leakage factor' in Theis' non-equilibrium equation: the 'ideally confined aquifer' was thus replaced by the 'multiple aquifer'. This concept was extended and refined by Neuman and Whitherspoon (1971, 1972), whose calculations illustrate clearly the time-dependent nature of the hydraulic behaviour of multi-aquifer systems. In pumped aquifer 1 of the two-aquifer system of Figure 3.28, the calculated drawdown at distance r from the well is similar for 'small values of time' to that of the Theis solution for an ideally confined stratum.

The deviation from the ideally confined behaviour increases, however, with increasing pumping time, as indicated by the 'intermediate values of time'

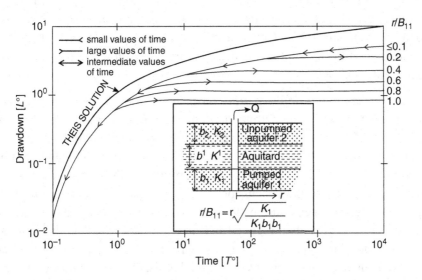

Fig. 3.28 Example of relative hydraulic behaviour of confined aquifer as function of time (Tóth, 1995, Fig. 1; simplified from Neuman and Witherspoon, 1971, Fig. IV-18).

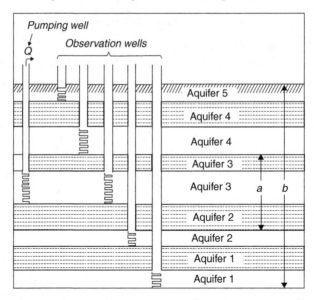

Fig. 3.29 Hydraulic continuity of multiple-aquifer systems 'a' and 'b', as inferred from short- and long-term pumping tests, respectively (Tóth, 1995, Fig. 2).

in Figure 3.28. And finally, all water is derived at 'large values of time' through leakage from the un-pumped surficial aquifer. If pumping were to stop early, the aquifer would be perceived as ideally confined; if only late drawdowns were considered, the aquifer would appear unconfined. These theoretical conclusions were empirically corroborated by a field test in which a time lag of approximately 30 days was observed across a shale bed 16 feet (4.9 m) in thickness (Fig. 3.24; Witherspoon and Neuman, 1967). Consequently, whether system 'a' or 'b' (Fig. 3.29) is deemed hydraulically continuous depends primarily on the relative lengths of pumping and observation.

(ii) Hydraulic continuity inferred from basin hydraulics. In what might be the earliest attempt at explaining petroleum accumulation by cross-formational water flow, Munn (1909) calls 'imperviousness of strata' a 'time-honored delusion' and continues: 'No unaltered sedimentary rocks of the type in which oil and gas pools occur are impervious' (Munn,1909, p. 516). From the 1950s on, regional studies have led increasingly to the conclusion that a quantitative interpretation of long pumping tests, regional pressure patterns, basinal water balances and large-scale flow models is possible only if regional and cross-formational hydraulic communication is assumed. Walton (1960) observed, for instance, that in the southern Illinois Basin 'Available data indicate that on a regional basis, the entire sequence of strata, from the top of the Galena-Platteville to the top of the shale beds of the Eau Claire Formation, essentially behave [sic] hydraulically as one aquifer' (Walton,1960, p. 18). Kolesov (1965) stated explicitly that thousands of observed pressures in

Siberia, Russia and the Dnieper-Donets basin are easily explained 'If the artesian waters in different strata are considered one single complex entity hydraulically connected' (Kolesov, 1965, p. 197). A group of French hydrogeologists (Albinet and Cottez, 1969; Astié *et al.*, 1969; Margat, 1969) interpreted vertical hydraulic head differences in various large sedimentary basins (e.g., Paris, W. Africa, Sahara, Aquitain) as indicating flow between confined aquifers through the intervening aquitards. They argued that if 'leakage' can be produced artificially by pumping, it must occur also under the effect of natural head differences. By means of a quantitative water-balance model of the Aquitain basin, France, covering an area of over 100 000 km^2, Besbes *et al.* (1976) verified the hydraulics of *'eight main aquifers, which communicate through aquitards'* (Fig. 3.30; Besbes *et al.*, 1976, p. 294).

More recently, Neuzil *et al.*, (1984) showed by numerical simulation that the actual amount and rate of water withdrawal from the Dakota aquifer would be impossible without leakage through the overlying Pierre Shale. The vertical permeability of the shales, however, must be 10–1000 times larger on the regional scale than on the local scales.

(iii) Convergence of transmissivities to regional field values inferred from pumping tests, Cretaceous, Alberta, Canada. Transmissivities determined by pumping tests of increasing lengths (Tóth, 1966b, 1973) approached a value evaluated from the annual pattern of regional water-level fluctuations (Tóth, 1968, 1982) in heterogeneous clastic rocks in the Alberta Sedimentary Basin, Canada (Fig. 3.31). The asymptotic convergence of transmissivities towards a limiting finite value suggests that the flow domain sampled by the water withdrawn through individual aquifers for a sufficiently long time is the same as the one traversed by the regional flow, i.e. that the rock framework is hydraulically continuous.

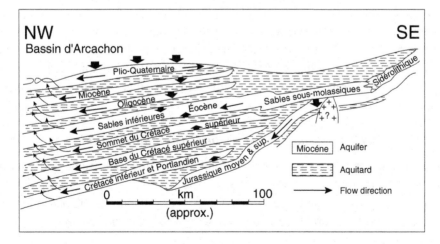

Fig. 3.30 Schematic of multiple-aquifer systems and computed flow paths, Aquitain Basin, France (Tóth, 1995, Fig. 3; modified from Besbes *et al.*, 1976).

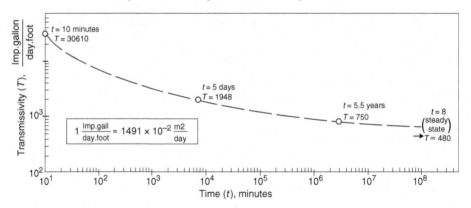

Fig. 3.31 Transmissivities determined from pumping tests of increasing duration and converging toward values inferred from seasonal water-level changes (Tóth, 1995, Fig. 4; compiled from Tóth, 1966b, 1968, 1973, 1982).

3.4.1.2 Additional arguments in support of regional hydraulic continuity

In addition to conclusions inferred from aquifer and basin hydraulics, regional hydraulic continuity is also supported by observations of rock-pore sizes and rock permeabilities.

Rock-pore sizes were estimated by Tissot and Welte (1978) to decrease to an apparent minimum of 1 nm at a depth of 4000 m, whereas the effective diameter of a water molecule is ≈0.32 nm only. Consequently, a water molecule should be able to pass through the intergranular voids of the tightest shales even if two monomolecular water layers are assumed to line the pore walls. Brace (1980) reported measurable permeability values in any rock type (*pure* halite being the only possible exception) and at any depth accessible by the days' drilling technology (to depths of 8–10 km). In the Swan Hills region of Alberta, Canada, minimum values for vertical hydraulic conductivities of massive clastic and evaporite aquitards were observed by Hitchon *et al.* (1989) to be about 7 to 9×10^{-11} m/s. Many other examples can be cited for extensive and tight rocks being effectively permeable, whereas not one reliably impermeable formation has yet been identified, to my knowledge, in the search for suitable repository sites for radioactive or other long-living hazardous wastes.

3.4.2 Consequences of regional hydraulic continuity

The natural consequences of hydraulic continuity of the rock framework can be summarized into three groups of processes and phenomena: (a) development of regionally extensive groundwater flow systems; (b) systematic areal distribution of matter and heat; (c) hydraulic interdependence of different basinal areas.

3.4.2.1 Regionally extensive groundwater flow systems

The basic form of regionally extensive steady-state flow systems facilitated by hydraulic continuity develops in the 'unit basin', as discussed in Chapter 2. To recall, the unit basin is a symmetrical topographic depression with linearly sloping flanks, a homogeneous rock framework, and with three distinctly different ground-water flow regimes: the recharge, midline, and discharge regions. These regions are characterized by descending, lateral and ascending flow, respectively, as well as by the associated patterns of pore pressure, hydraulic head and vertical pressure gradients (Fig. 2.2). The closest natural equivalents of the unit basin are inter-montane valleys and closed lake basins (e.g., Issar and Rosenthal, 1968; Mifflin, 1968; Ortega and Farvolden, 1989; Tóth and Otto 1993). In regions with substan-tial local topography, such as high-relief platforms and foreland basins, *composite flow patterns* develop. These patterns are topographically modified versions of the unit basin flow systems. In such basins three types of flow systems are recognized: local, intermediate and regional. Composite patterns are characterized by laterally alternating recharge–discharge regions, contiguity of hydraulically similar regions of flow systems of different order, vertical superposition of different types of flow regimes, and stagnation points (Fig. 3.3; e.g., Astié et al., 1969; Erdélyi, 1976; Tóth and Almási, 2001).

Low permeability strata may modify basinal pore-pressure patterns to such a degree that the erroneous impression of hydraulic discontinuity is created. The apparent discontinuity suggested by the $p(d)$-profiles in Figure 3.11 is due to a lack of pressure measurements in the aquitard, rather than to a lack of flow through it. A hiatus in observed pressure values can lead to similarly erroneous conclusions regarding hydraulic continuity in flow domains of non-gravitational systems also. Such can be the case, for instance, in systems of expelled water of compacting sedimentary basins, such as the Gulf of Mexico.

Owing to hydraulic continuity, basinal pore pressures adjust to changing bound-ary conditions albeit, perhaps, with a time lag. The process may result in *heterochronous flow fields* (Fig. 3.26). Such adjustment was estimated to take approximately 4 Ma through a 480 m thick shale stratum in northern Alberta, (Fig. 3.25b; Tóth and Millar, 1983).

3.4.2.2 Systematic distribution of matter and heat: the geologic agency of groundwater

Moving groundwater interacts with its environment through a variety of physi-cal, chemical and mechanical processes. Groundwater mobilizes, transports and deposits matter and heat, modifies pore pressures, and lubricates fracture plains and soil grains. The effects of these processes are different, even opposite, in high-energy recharge and low-energy discharge areas. Oxidation, dissolution, moisture deficiency, removal of matter, subhydrostatic pressures and reduced subsurface temperatures are common in the former, whereas the opposite conditions characterize the latter (Tóth, 1984, 1999).

An important aspect of the geologic agency of groundwater is that its manifestations occur in any size and any order of flow systems, as well as at any depth and any time scale. The diversity of groundwater-generated field features is increased by the effects of local environments. For instance, just a simple excess of water in a discharge area can result in springs, seeps, soil creep, quick sand, ice mounds, fens and so on, depending on the conditions of climate, soils and topography. In the deeper subsurface and on the scale of geologic time, hydraulic continuity facilitates the development of extensive flow systems. In combination with other requisite conditions, such flow systems may become effective transport mechanisms in the generation of various types of metallic ore deposits (Galloway and Hobday, 1983; Baskov, 1987; Garven *et al.*, 1993); petroleum accumulations (Tóth, 1988; Wells, 1988; Verweij, 2003); and geothermal heat anomalies (Smith and Chapman, 1983; Beck *et al.* 1989; Lazear, 2006).

3.4.2.3 Hydraulic interdependence of different basinal regions and hydrologic components

An important consequence of hydraulic continuity is the propagation of hydraulic stresses imposed on the groundwater body to distant geographic regions, to unexpected depths and strata, for great lengths of time, or in unexpected forms. As an example, Figure 3.32 presents the theoretical response of various water-balance components to pumping in a regionally unconfined basin. The effects most relevant in the present context are the unexpected decline in natural recharge at time $= t_4$, and the conversion of former discharge areas into recharge areas (between t_3 and t_4) upon extended pumping.

In the case of a real-life example, pumping for the city of Tokyo in the Kanto groundwater basin has affected groundwater levels in a radius of over 100 km and it has caused land subsidence in excess of 4 m in places (Nirei and Furuno, 1986) In the design of repositories for nuclear wastes, changes in pore pressures must be considered for thousands of years after possible changes in boundary conditions, and oil fields may be remigrated by changes in natural flow-field boundaries (Fig. 3.26; Tuan and Chao, 1968), or by pumping in apparently unconnected locations (Hubbert, 1953).

3.4.3 Conclusions

Hydrogeologically, a large sedimentary basin can be envisaged as an extensive structural depression of the Earth's crust filled with sedimentary rock and possibly

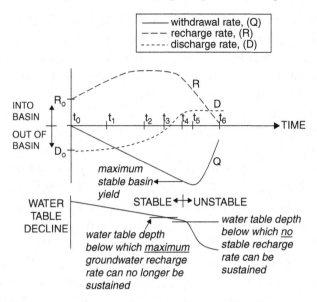

Fig. 3.32 Schematic diagram of transient relations between rates of recharge, discharge and water withdrawal in a regionally unconfined drainage basin (Tóth, 1995, Fig. 11, from Freeze, 1971, Fig. 11).

containing several drainage basins. Owing partly to large contrasts in permeabilities, sharply differing groundwater conditions can develop in different regions and at different depths. Water flow may be generated by gradients of different energy types, e.g., gravity, pressure, thermal and chemical.

Significant or abrupt variations of groundwater conditions in large sedimentary basins have often been attributed to hydraulic discontinuity between different parts of the rock framework. Increasingly, however, this thinking is giving way to the perception that the rock framework is hydraulically continuous. In this view a subsurface rock body is hydraulically continuous on a given time scale if a change in hydraulic head (or pore pressure) at any point can cause a head change at any of its other points within a time interval measurable on the specified time scale. This phenomenological property is, therefore, dependent on the chosen scales of space and time.

The most fundamental consequence of regional 'leakage' is that large-scale flow systems develop that are adjusted, or are in the process of adjustment, to fluid-potential boundaries controlled by maximum and minimum values. These values may be due to topographic highs and lows, extensive surface-water bodies, compressional or compactional pressure maxima, or dilatational pressure troughs. The flow systems themselves function as conveyor belts and entail systematic distributions of heat and matter within the basins. They thus make groundwater an effective and quasi-ubiquitous geologic agent.

Groundwater is active simultaneously on different scales of space and time. The effects of moving groundwater are varied and environmentally modified. They include, for instance, regionally contrasting soil-moisture conditions; various soil and rock mechanical phenomena, such as liquefaction and land slides; wetland development; effects on the base flow of rivers; regular patterns of groundwater chemistry; diagenetic changes of minerals; soil salinization; migration and accumulation of various metallic minerals and petroleum; and the generation of certain physical and chemical signatures that indicate the presence of those accumulations.

Perhaps the most challenging, and rewarding, task of the hydrogeologist is the selection of the appropriate scales of space and time for a particular problem. For example, the hydraulic engineer may easily overestimate the sustainable yield of a good aquifer by not recognizing that the ultimate constraint on its production is the bulk water-transmitting ability of the entire rock body of the basin, or even the annual precipitation. Conversely, pumpage required to keep an open-cast mine dry might be underestimated based solely on the pumped aquifer's transmissivity and storativity, i.e. if leakage through apparently impervious aquitards (which could be recognized by regional flow studies) is ignored. Or, the same boundary, which the petroleum-reservoir engineer rightly assumes for the purpose of production-rate calculations to be impermeable to hydrocarbons and to envelop a lenticular accumulation, may allow the passage of sufficient hydrocarbons on the time scale of petroleum migration for the lens to be targeted by the explorationist as a prospective play.

In summary, the correct interpretation and utilization of the great number of groundwater-related processes and phenomena as well as the correct modelling and prediction of the effects of stresses imposed upon the groundwater regime in large sedimentary basins, thus the correct practice of hydrogeology, require the recognition of that fundamental property of the rock framework: hydraulic continuity.

4

Gravity flow of groundwater: a geologic agent

4.1 Introduction

Chapter 4 is intended to advance the view that moving groundwater is the common basic cause of a wide variety of natural processes and phenomena and hence it should be regarded as a general geologic agent.

That groundwater plays an active role in certain geologic processes has been recognized in numerous earth-science subdisciplines for a long time. However, the generality of this role was not appreciated until the 1960s and 1970s, when the underlying common cause itself was sufficiently understood to allow, and indeed to stimulate, dedicated studies of its broader ramifications. That cause was the gravity-driven basinal flow of groundwater. Even during this period, however, the generalization of groundwater's role in nature was hindered by at least two factors. First, the diversity of natural phenomena related to groundwater flow effectively conceals a single common cause. Second, a lack of knowledge, or even awareness, of basinal groundwater hydraulics among specialists of various other disciplines prevents them from recognizing the cause-and-effect relation between regional groundwater flow and the particular phenomena of their interest. By way of illustrating the difficulty of envisaging a common origin, suffice only to list such diverse, and indeed in some cases disparate, natural phenomena generated and/or fundamentally shaped by groundwater flow as: soil salinization, continental salt deposits, regional patterns of groundwater's chemical composition, soil liquefaction, gullying, landslides, dry-land ice fields, geysers, positive and negative geothermal anomalies, lake eutrophication, base-flow characteristics of streams, bog- vs. fen-type wetlands, type and quality of plant species and associations, taliks in permafrost, roll-front and tabular uranium deposits, dolomitization of limestones, karst morphology, diagenesis of certain clay minerals, some sulfide-ore deposits, and certain types of hydrocarbon accumulations.

Examples of studies, including some major papers and books, that treat of specific geologic processes while recognizing the role played by groundwater include: Yaalon (1963), Back (1966), Zaruba and Mencl (1969), Williams (1970), Deere and Patton (1971), Domenico and Palciauscas (1973), de Vries (1974), Boelter and Verry (1977), Winter (1978), Gerrard (1981), Wallick (1981), Fogg and Kreitler (1982), LaFleur (1984), Garven (1989), Macumber (1991), Stute *et al.* (1992), Garven *et al.* (1993), Stuyfzand (1993), Cardenas (2007) and Wörman *et al.* (2007).

On the other hand, attempts at advancing the notion of groundwater as a general geologic agent by focusing on its multifarious effects in nature are exemplified by: Tóth (1966a, 1971, 1972, and 1999), Freeze and Cherry (1979), Engelen and Jones (1986), Deutscher Verband für Wasserwirtschaft und Kulturbau e.V. (1987), Back *et al.* (1988), Ortega and Farvolden (1989), Engelen and Kloosterman (1996), Domenico and Schwartz (1997) and Ingebritsen and Sanford (1998).

The recognition of moving groundwater as a common cause of such a diversity of geologic effects was not possible until the system-nature of its regional flow distribution had been understood. The period of conscious efforts to model, observe, and evaluate basin-scale flow of groundwater and the factors controlling it began in the early 1960s. Initially, attention was directed to small basins and gravity-driven flow (Tóth, 1962a, 1963; Freeze and Witherspoon, 1966, 1967, 1968). The essential results, that are relevant also in the present context, were the recognition that in topography-controlled flow regimes, groundwater moves in systems of predictable patterns and that various identifiable natural phenomena are regularly associated with different segments of the flow systems (Mifflin, 1968; Williams, 1968, 1970; Astié *et al.*, 1969; Freeze, 1969).

The scope of investigations was expanded later to extensive and deep basins. These studies have revealed or confirmed: (i) the hydraulically continuous nature of the rock framework, which facilitates large-scale cross-formational flow systems; (ii) time lags in flow-pattern adjustments to changing boundary conditions, ranging from human to geological time scales; (iii) the multiplicity of possible flow-inducing energy sources. It became obvious also that, although in many respects different from their small-basin counterparts, a wide variety of geologic phenomena is associated with large-scale flow systems (Bredehoeft and Hanshaw, 1968; Neuman and Witherspoon, 1971; Erdélyi, 1976; Tóth, 1978; Neuzil and Pollock, 1983; Tóth and Millar, 1983; Neuzil *et al.*, 1984; Bethke, 1985; Tóth and Corbet, 1986; Goff and Williams, 1987; Parnell, 1994).

The recognition that basinal flow of groundwater occurs in systems on a wide range of spatial and temporal scales has provided a unifying theoretical background for the study and understanding of many seemingly unrelated natural processes and phenomena. It has shown the gravity flow of groundwater to be a general geologic agent. The thesis is demonstrated through an overview of the basic causes, principal processes, and various natural manifestations of regional groundwater flow.

4.2 The basic causes

Two fundamental causes make gravity-driven groundwater a geologic agent: (i) *in-situ* interaction between water and its ambient environment; (ii) transport by flow organized into hierarchical systems of different magnitude. The interaction of water with its surroundings generates various natural processes, products and conditions. Systematized flow paths, on the other hand, function as sustained mechanisms of distribution of those effects into regular spatial patterns within the basinal flow domain. In basins where groundwater flow is controlled by the topography, the spatial distribution patterns of groundwater's effects are functionally related to identifiable and characteristic segments of the flow systems. Such relations make correlation between cause (groundwater flow) and effect (natural processes, conditions and phenomena) feasible and verifiable. This is the chief argument for limiting the scope of this discussion to gravity flow.

4.2.1 In-situ *interaction between groundwater and its environment*

Interaction between groundwater and its natural environment can occur in different forms and is driven by the various components and attributes of the two systems seeking equilibration. For this overview, three main types of interaction are identified between groundwater and its hydrogeologic environment, namely: chemical, physical and kinetic. The tendency toward equilibration between different systems (i.e. the seeking of a state of minimum free energy, whether they are chemical, mechanical, kinetic or thermal) is a basic law of nature and is accomplished through various natural processes. As a result of such natural processes (reviewed below), water moving through the subsurface can: (i) mobilize and deposit matter and heat; (ii) transport matter and heat; (iii) lubricate discontinuity surfaces in the rock framework; (iv) generate and modify pore pressures.

Groundwater flow can thus produce various *in-situ* effects, the nature of which depends on the chemical, physical and hydro-kinetic conditions of a particular locality. In areas of high chemical and thermal energy, minerals are added to the water by dissolution, oxidation, attack by acids and other processes. Relatively high mechanical energy causes the water possibly carrying dissolved mineral matter and heat to move away from a site, thus rendering the locality a supply source of minerals, water and heat. Conversely, in regions of low chemical, thermal and kinetic energy, the water tends to converge and possibly leave the subsurface domain by discharging on to the land surface, stream bed, or lake bottom; precipitate mineral matter; and lose heat. Collectively, the diverse manifestations resulting from the interaction of groundwater with its environment at a given locality may be called the *in-situ* environmental effects of groundwater.

4.2.2 Flow: a mechanism of systematic transport and distribution

In-situ environmental processes alone would not normally be sufficient to render groundwater a significant geologic agent, because most of them are self-limiting in time and/or randomly scattered in space. For most geologic phenomena to develop, such as mineral deposits, geothermal anomalies, wetlands, and so on, it is necessary that the products of disequilibria accumulate over sufficiently long periods of time. Such accumulations usually occur by concentration within relatively small rock volumes or in areas supplied from more extensive source regions. Other phenomena, such as vegetation patterns, soil and rock mechanical instability, ice development, etc., are secondary effects consequent upon the primary processes.

The only mechanism capable of producing and maintaining the required disequilibrium conditions for such a wide range of natural phenomena is the regional flow of groundwater. Individual flow systems may be thought of as subsurface conveyor belts, with their source regions being the areas of mobilization and loading, and their terminal regions the areas of delivery and deposition. The middle segments function chiefly as relatively stable environments of mass and energy transfer.

A schematic overview of groundwater flow distribution and some typical hydrogeologic conditions and natural phenomena associated with it in a gravity-flow environment is presented in Figure 4.1. The figure shows an idealized basin with one single flow system on the left-side flank of insignificant local relief, and a hierarchical set of local, intermediate and regional flow systems of the right flank of composite topography. Each flow system, regardless of its hierarchical rank, has an area of origin or recharge, an area of throughflow or transfer, and one of termination or discharge (see Chapter 3.1). In the recharge areas, the hydraulic heads, representing the water's mechanical energy, are relatively high and decrease with increasing depth; water flow is downward and divergent. In discharge areas, the energy and flow conditions are reversed: hydraulic heads are low and increase downward, resulting in converging and ascending water flow. In the areas of throughflow, the water's mechanical energy is largely invariant with depth (the isolines of hydraulic heads are subvertical) and, consequently, flow is chiefly lateral. The flow systems operate as conveyor belts that effectively interact with their ambient environment. The interaction produces *in-situ* environmental effects and the flow serves as the mechanism for mobilization, transport (distribution) and accumulation.

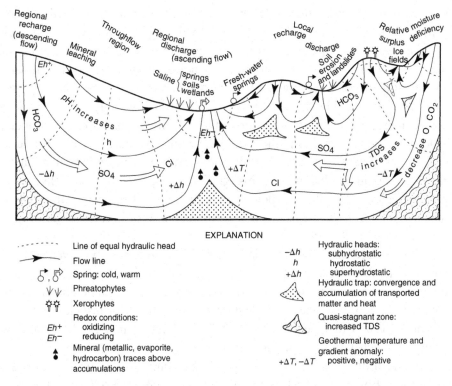

Fig. 4.1 Effects and manifestations of gravity-driven flow in a regionally unconfined drainage basin (Tóth, 1999, Fig. 1, modified from Tóth, 1980, Fig. 10).

Typical environmental effects and conditions due to gravity-flow systems of
groundwater are illustrated in Figure 4.1 and include: (1) subhydrostatic, hydrostatic, and superhydrostatic hydraulic heads at depth in the direction of flow from
recharge, to discharge areas, respectively; (2) relatively dry surface-water and soil-
moisture conditions (negative water balance) in recharge areas, and water surplus
(positive water balance), possibly resulting in wetlands, in discharge areas (these
conditions are expressed in comparison with an average water balance in the basin,
which would result solely from precipitation and evapotranspiration on a horizontal tract without surface runoff); (3) systematic changes in the water's anion
facies, from HCO_3 through SO_4 to Cl, both along flow systems and with depth;
(4) chemically leached soils and near-surface rocks in areas of inflow, but increased
salt contents possibly amounting to salt-affected soils or even commercial salt
deposits at flow-system terminuses; (5) saline marshes in situations where wetland
conditions and intensive salt supply coincide; (6) negative and positive anomalies of geothermal heat and geothermal gradients beneath areas of descending and
ascending flows, respectively; (7) chemically oxidizing and reducing conditions

in the near-surface environment of recharge and discharge areas, respectively; (8) identifiable response in the type and quality of the vegetation cover to the contrasting nutrient and moisture conditions generated by the inflow and outflow of groundwater at flow-system extremities; (9) increased vulnerability of the land surface to soil and rock mechanical weakness failures (e.g., soil erosion, slumping, quick grounds and landslides) in areas of discharge, possibly developing into major geomorphologic features as, for instance, gullying and stream meanders; (10) accumulation of transported mineral matter such as metallic ions (iron, uranium, sulfides of various metals), hydrocarbons, and anthropogenic contaminants, primarily in regions of flow paths converging from opposite directions (hydraulic traps) or in regions where the fluid potential is minimum with respect to a transported immiscible fluid (oil, gas), e.g., at grain-size boundaries or in rocks of adsorptive minerals.

The above effects and conditions have been attributed to gravity-driven systems of groundwater flow. However, similar natural effects can be generated by flow due to other sources of energy, such as: sediment compaction; tectonic compression; thermal convection; buoyancy; dehydration of minerals as, for instance, gypsum; and osmosis. The history, geometry and geologic effects of non-gravitational flow systems are much less tractable and, therefore, not yet amenable to systematic and generalized discussions. Some outstanding examples of particular cases are represented by Berry (1973), Oliver (1986), Bethke (1989), Hanor (1987), Harrison and Summa (1991) and Tóth and Almási (2001).

4.2.3 Ubiquity and simultaneity

The key reason for such diversity in the geologic effects of groundwater is that its flow systems are active over broad spectra of space and time. Water is present in the entire porous section of the Earth's upper crust possibly penetrating by cross-formational communication to depths of 15–20 km. Throughout this depth range water is in continual motion although at vastly differing rates, possibly ranging from more than 10^{-3} m/s near the surface to less than 10^{-12}–10^{-14} m/s at depth. Also, the stability of individual flow systems is a function of depth and/or lateral position within the flow domain. Flow paths near the land surface and/or in unconfined domains adjust to changes in boundary conditions more readily than at great depths or in confined situations (see Section 3.1.3). Geologic effects of flow that require long-term hydraulic stability to be produced (e.g., mineral or petroleum accumulations) are therefore less likely to fully develop at shallow depths (Fig. 3.26).

Also groundwater interacts with its environment simultaneously at all depths and at all times. However, the physical, chemical and thermal conditions, the reactive matter, the processes and the flow rates prevailing at different locations of the land surface and at different depths of the flow regimes are vastly different. As a result, the rates and products of interaction between water and its environment can be fundamentally different though generated simultaneously at different places, but by the same agent: moving groundwater.

4.3 The main processes

Three types of disequilibrium conditions were identified in the previous section as the primary causes for groundwater's geologic agency: chemical, physical and kinetic. The disequilibria tend to be reduced in nature through numerous and different processes, of which a brief, annotated enumeration is presented in this section. Their detailed discussion is not warranted here as they are all processes and material interactions dealt with by various works in the present context (Freeze and Cherry, 1979; Fetter, 1994; Schwartz and Zhang, 2003), and they are also well known from basic science.

4.3.1 Chemical processes

Dissolution is one of the most generally effective processes in groundwater chemistry and it could be considered as the first step in the water's chemical evolution. It affects both gases and solids. Typical regions of gas dissolution by groundwater are the soil zone, unsaturated or vadose zone, and zones of oil and gas accumulations. Gases commonly interacting with groundwater include N_2, Ar, O_2, H_2, He, CO_2, NH_3, CH_4 and H_2S. Dissolution of gases may render the water weakly acidic, i.e. chemically aggressive. The degree of dissolution depends on the solubility of the minerals, the antecedent concentration of the water, and the pressure and temperature of the locality. Rocks that are most soluble include limestone ($CaCO_3$), dolomite ($CaMgCO_3$), gypsum ($CaSO_4$), halite (NaCl) and sylvite (KCl); these rocks serve also as the principal source for Ca, Mg, Na, K, CO_3, SO_4 and Cl in groundwater. Silicates and other so-called 'insoluble rocks' are also soluble to some degree under certain chemical or/and thermal conditions and yield the minor or trace constituents in groundwater.

Hydration is the penetration of water into the crystal lattice of minerals or the attachment of water molecules to the ions of dissolved salts. It is a common process in the evolution of groundwater's chemical composition and plays an important part, among others, as the first step in the weathering of minerals such as, for instance, anhydrite to gypsum or biotite to vermiculite.

Hydrolysis is defined as the reaction of any substance with water. More strictly, hydrolysis is the reaction of an ion with water to form an associated species plus H^+ or OH^-. Hydrolysis of cations produces slightly acidic solutions according to the general equation $M^+ + H_2O \rightleftharpoons MOH + H^+$, whereas anion hydrolysis results in a basic solution $X^- + H_2O \rightleftharpoons HX + OH^-$. Hydrolysis is an effective process only if (1) some of the produced ions are removed from the solution, or (2) others that are required for the hydrolysis are added to the solution. Otherwise, chemical equilibrium, thus cessation of dissolution, occurs.

Oxidation–reduction reactions are chemical reactions in which electrons are transferred from one atom to another. The process of oxidation is the result of a loss of free electrons by the substance being oxidized, whereas reduction is the gain of free electrons by the reduced substance. Oxidation and reduction must occur together and also compensate each other. Oxidation, as a process of modifying water quality, is most important in the vadose zone, where there is a copious supply of O_2 from the air and from CO_2. Below the water table in the saturated zone, water is rapidly depleted in O_2. Also, the solubility of O_2 in water is low in contrast to its solubility in air (6.6 and 200 cm^3/l at 20°C, respectively). Consequently, the importance of oxidation also decreases rapidly with depth. Typical and important oxidation processes include the oxidation of sulfides, producing Fe_2O_3, H_2SO_4 and CO_2, with the acids attacking carbonates; magnetite, producing limonite; and organic matter such as lignite, coal and bitumen, producing CO_2. Reduction also is important in organic deposits, which constitute reducing chemical environments. In these cases, oxygen may be obtained from oxides, sulfates and nitrates as well as nitrites and from various gases. As a result, ionic species may be generated, including H_2, H_2S, CH_4 and other hydrocarbons, and S^{2-}, NO_2^-, NH, Fe^{2+} and Mn^{2+}. Some metallic compounds, such as UO_2 (pitchblende), are chemically stable in reducing environments and may be trapped there to form deposits.

The most commonly occurring acids contributing to the chemical evolution of groundwater by direct attack on the rock framework include carbon dioxide (CO_2), nitric acid (NO_2), sulfuric acid (H_2SO_4) and various organic acids (humic, fulvic).

Chemical precipitation of mineral matter dissolved in groundwater may occur for several reasons, including: (1) reaction with ions from the solid framework to form insoluble precipitates, such as the formation of fluorspar (CaF_2) by the reaction of fluoride in water with calcium of the country rock; (2) changes in pressure and temperature affecting the solubility of certain chemical constituents in water, as in the precipitation of calcareous tufa ($CaCO_3$) around the orifices of springs due to the release of dissolved CO_2 on pressure decrease, or the deposition of silica (SiO_2) from hot springs owing to a drop in temperature; (3) oxidation of dissolved matter exposed to air resulting in components of decreased solubility such as the precipitation of ferric hydroxide ($Fe(OH)_3$) from waters containing ferrous iron ($Fe(OH)_2$) in solution.

Base exchange, or ion exchange, is the process in which ions and molecules adsorbed on the surfaces of solid substances by physical or chemical forces (van der Waals attraction and 'chemisorption', respectively) are exchanged for ions in the water. The most important substances capable of ion exchange are clay minerals, such as kaolinite, montmorillonite, illite, chlorite, vermiculite and zeolite, ferric oxide, and organic matter, mainly because they form colloids of large surface areas. An important example for base exchange is the replacement of Na by Ca and/or Mg in bentonite resulting in a natural softening of the water by enrichment in Na, on the one hand, and increased porosity and permeability of the newly-formed Ca- and/or Mg-based clay mineral, on the other.

Sulfate reduction is due mainly to bacteria and contact with organic matter (coal, lignite, petroleum) and results in the removal of sulfates from the transporting groundwater. One example is the reaction of sulfate water in contact with methane:

$$CaSO_4 + CH_4 \rightleftharpoons CaS + CO_2 + 2H_2O;$$

$$CaS + 2CO_2 + 2H_2O \rightleftharpoons H_2S + Ca(HCO_3)_2.$$

Concentration of the total dissolved solids (TDS) content in groundwater may be effected also by evaporation of water and solution of mineral matter, in addition to various chemical reactions indicated earlier. Concentration by evaporation is operative mainly in the soil-moisture zone and between rainfall events. In this zone, owing to evaporation of earlier rainwater, salt concentration increases and the dissolved salts are washed down to the water table by subsequent rains. The higher the temperatures and the longer the periods between rainfall events, the higher is the degree of salt concentration in the groundwater. On the discharge ends of flow systems, concentration by evaporation may lead to soil salinization and/or the formation of continental salt deposits as the transported salts precipitate from the groundwater upon emerging at the land surface (Williams, 1970; Tóth, 1971; Wallick, 1981; Jankowski and Jacobson, 1989; Macumber, 1991).

Concentration by solution may also be an important hydrochemical process in the subsurface. The maximum concentration attainable by this process is limited by the chemical equilibrium between the rock and water. If, however, there is a sufficiently long time for the water to dissolve mineral matter, a very small amount of soluble minerals in the rocks will result in relatively high water salinity. For instance, in leached clays with a NaCl concentration of 1–2%, a density of 2.2, and porosity 40%, water in equilibrium with the rock will contain 13.2–26.4 g/l of NaCl.

Ultrafiltration by shale membranes is thought to be a possible mechanism for increasing ionic concentrations in groundwater. The hypothesis postulates that compacted clays and shales may perform as imperfect semipermeable membranes through which water molecules are forced by differences in hydraulic heads. The ions of chemical species are left behind thus increasing the salt concentration on the high-pressure side of the argillaceous beds. The process is probably not as prevalent and general in nature as its original proponent thought (Berry, 1969).

4.3.2 Physical processes

Lubrication of discontinuity boundaries by water in the rock framework, such as grain surfaces in soils and in unconsolidated sediments or fracture and fault planes in indurated rocks, reduces friction and enhances the effectiveness of shear stresses, possibly acting upon the discontinuities. As a consequence, shear movements of soil and rock material can be induced along the discontinuities, ranging in magnitude from minor rearrangement of mineral grains, as in compaction, to major land slides and earth quakes. The process is particularly effective in regions where large variations in precipitation cause large fluctuations in the water table (Deere and Patton, 1971; Bonzanigo *et al.*, 2007). The effect is further magnified by high or increased pore pressures (neutral stresses) that help to reduce frictional resistance to shear displacement. Relatively high or increased pore pressures are natural attributes to the discharge segments of groundwater flow systems.

Pore-pressure modification, relative to hydrostatic values, can have various causes, with widely ranging geological consequences. Reductions or increases in pore pressures, with respect to either space or time, affect the types, rates, and directions of chemical reactions, the solubility of gases, gas-saturation levels in subsurface fluids, and the strength thus deformation, and even the integrity of rocks (Hubbert and Rubey, 1959; Freeze and Cherry, 1979; Gretener, 1981; Tóth and Millar, 1983; Domenico and Schwartz, 1997).

Modification of pore pressures by gravity-driven flow systems is a dynamic effect. It changes systematically from negative through zero to positive in recharge, throughflow, and discharge areas, respectively (Fig. 4.1; Tóth, 1978). Based on Terzaghi's (1925) theory of stress relations in saturated rocks, these changes lead to increases in effective stresses, i.e. to increased strength, in soils and unconsolidated rocks in recharge areas, while reducing the effective stress, i.e. the strength, at a system's discharge end. The result is an increased vulnerability of the land surface to soil erosion, landslides, and other forms of mass wasting in discharge regions (Miller and Sias, 1998; Bonzanigo *et al.*, 2007; Eberhardt *et al.*, 2007) and a relative increase in the stability of hill slopes, river banks and various positive geomorphologic features in areas of recharge (Iverson and Reid, 1992; Reid and Iverson, 1992).

4.3.3 Kinetic or transport processes

The *transport* of water itself in flow systems is groundwater's most fundamental process in playing its role as a geologic agent. In addition to its functions as a medium of transport and a reactive agent in the subsurface environment, water is also a component of that environment as well as of the hydrosphere of the land surface. By means of flow systems, the water masses are regularly distributed in a gravitational flow domain and thus create regionally contrasting moisture conditions between the regions of inflow and outflow. The process may be viewed as the subsurface portion of the hydrologic cycle. As such, it is a key element in the areal distribution of water bodies, their chemical characteristics, the type and magnitude of water-table fluctuations, and the many far-reaching ecological ramifications linked to these conditions (Klijn and Witte, 1999; Batelaan *et al.*, 2003).

The *transport of aqueous and non-aqueous matter* in and by groundwater flow systems also is fundamental among the subsurface geologic processes. A wide variety of matter is transported by groundwater in many different forms, including aqueous solutions of organic and inorganic ions; particulate matter in colloidal form or larger sized suspended grains; gases as bubbles or in solution; globules, micelles or ionic solutions of liquid hydrocarbons; and viruses and bacteria. The importance of transport of these matters by groundwater derives from the cumulative effects of the sustained processes resulting in the leaching and removing of minerals from soils and rocks, carrying nutrients to surface-water bodies, building and remigrating deposits of metallic and non-metallic minerals and hydrocarbons, causing water washing and biodegradation of ore deposits and oil accumulations, and concentrating contaminants at fluid-dynamically suitable subsurface locations.

Heat transport by moving groundwater is one of the most visible and most well understood geologic processes in the subsurface (Schoeller, 1962; Smith and Chapman, 1983; Romijn *et al.*, 1985; Rybach, 1985; Beck *et al.*, 1989; Deming, 2002; Anderson, 2005; Lazear, 2006). Water can contain, and thus transport heat because of its specific heat capacity, $\rho_w C_w$, where ρ_w is the water's density and C_w is its specific heat. If thermal disequilibrium i.e. a temperature difference exists, between the water and its ambient environment, heat is induced to flow in the direction of lower temperature. The rate of heat transfer depends on numerous factors, including the thermodynamic properties of the saturated rock framework and the water, the temperature difference between the water and its environment, and the rate of the water flow. The greater the rate of heat exchange the faster the temperature difference (the temperature anomaly) is reduced between the water and the environment.

The rate of heat exchange between moving groundwater and the rock framework is essentially the result of a competition between the dissipation of heat by conduction and the advective removal of heat by the flow of the water. This competition is expressed quantitatively by the Péclet number: $P_e = \frac{n\rho_w C_w}{(\rho_m C_m)} \frac{qD}{k_m}$, where n is porosity of the rock; ρ_w, ρ_m are density of water and saturated medium, respectively; q is specific volume discharge; D is flow-path length; $k_m = \lambda_m / \rho_m C_m$ is thermal diffusivity, in which λ_m is thermal conductivity and $\rho_m C_m$ is the specific-heat capacity of the saturated porous medium. In cases where $P_e < 1$, for instance due to slow flow, dissipation of heat by conduction dominates and temperature differences between water and rock equilibrate. If, however, $P_e > 1$, then advection, i.e. transport of heat by moving groundwater, prevails over the tendency for equilibration, and temperature anomalies are created and/or maintained. The process is enhanced by highly permeable discontinuities such as faults, fracture zones, cavities and solution channels, which are often major attributes in the formation of hot springs, hydrothermal ore deposits or other types of geothermal anomalies.

4.4 Manifestations

The geologic agency of moving groundwater is most plausibly revealed by the numerous and diverse effects and manifestations resulting from the various flow-driven chemical, physical and kinetic processes. Nonetheless, in spite of their collectively great number and diversity, groundwater-induced natural phenomena comprise a relatively small number of basic types, with variations within each type being due to site-specific characteristics of the 'hydrogeologic environment'.

4.4.1 The hydrogeologic environment

The 'hydrogeologic environment' is a conceptual system of those morphologic, geologic and climatologic parameters that determine the principal attributes of the 'groundwater regime' in a given area (Tóth, 1970). The six main attributes, or parameters, of the groundwater regime are: (1) water content in the rocks; (2) geometry of the flow systems; (3) specific volume discharge; (4) chemical composition of the water; (5) temperature; (6) the variations of all these parameters with respect to time.

The parameters of the groundwater regime are controlled by the three components of the hydrogeologic environment, namely: (1) the topography; (2) geology; (3) climate. In turn, the components are made up of various parameters. For topography, such parameters are: the size and shape of depressions, slopes, and prominences of the land surface, and the orientation and frequency of geomorphic features. For geology: soluble mineral content; type, nature, and geometry of rock bodies of different permeabilities, e.g., stratification, lenticularity, faulting, fractures, karst, and degree of anisotropy. For climate: temperature; amount, type, and seasonal variation of precipitation, and potential evaporation.

The controlling effect of the hydrogeologic environment on the groundwater regime is easily seen when the roles of the individual components are considered. The climate determines the amounts and distribution of water supplied to any region. Topography provides the energy available to the water to move, i.e. it determines the distribution of the flow-inducing energy by shaping the boundaries of the flow domain. Finally, geology constitutes the receptacle and conduit-system which control the possible amounts, rates, and local patterns of flow, as well as the distribution and amount of stored water. Geology determines the chemical constituents also and it may contribute local or regional sources of energy to the flow field by, for instance, compaction, compression or heat.

In nature, the various environmental parameters can combine in a virtually unlimited number of ways, each of them modifying a groundwater-induced basic process or phenomenon in a different way. For instance, soil-mechanical effects of discharging groundwater may be manifest by simply liquefaction (quick sand, quick clay) in a flat, plains region; by soil creep, slumping, and landslides in hills and mountains; or ice mounds and frost heaving under cold climates. Spring water discharging at the end of a simple flow system may be cold and fresh, hot or saline, depending on whether the system is shallow in a cold climate, passes by a subsurface heat source, or across beds of evaporites. Trees, characteristic of groundwater discharge areas are, among others, willow, alder and tamarack in cold climates; birch, poplar and oak in temperate regions; and palms in dry tropics.

4.4.2 Types of manifestations

In order to facilitate an easier overview of the multitude of effects and manifestations generated by gravity-driven groundwater flow and modified by environmental conditions, they have been grouped into six basic types with some sub-categories identified as follows (Tóth, 1984).

(a) Hydrological and hydraulic.
 (i) Local water balance; (ii) regionally contrasting moisture conditions; (iii) water-level fluctuations.
(b) Chemical and mineralogical.
 (i) Areal patterns of water salinity and isotope distribution; (ii) soil salinization and continental salt deposits; (iii) weathering, dissolution, and cementation; (iv) diagenesis of minerals.
(c) Botanical.
 (i) Species of plants; (ii) quality of plants.
(d) Soil and rock mechanical.
 (i) Liquefaction; (ii) soil erosion; (iii) slope stability.

(e) Geomorphological.
   (i) Erodability and stream valleys; (ii) karst development; (iii) geysers and mud volcanoes; (iv) frost mounds, pingos, ice fields.
(f) Transport and accumulation.
   (i) Temperature distribution patterns; (ii) low-temperature sulfide ores; (iii) uranium deposits; (iv) hydrocarbon fields, methane halos, oil seeps; (v) effluents and contaminants.

The common, and in most cases principal, factor producing all phenomena in these categories is the effect of groundwater itself. Several levels of further divisions can be formed in each case until site-specific effects and manifestations of groundwater flow are identified as shaped by the micro-environments of particular localities. By way of illustration, some examples are presented for each of the main types of natural phenomena below.

4.4.2.1 Hydrology and hydraulics

Two important hydraulic phenomena associated with gravity-induced basinal flow of groundwater are the areally systematic orientation of the flow's vertical components, and the relief-dependent depth of the phreatic belt (annual fluctuation of the water table). Flow is generally descending under positive morphological features, whereas it is ascending beneath topographic depressions (Figs. 4.1 and 4.2).

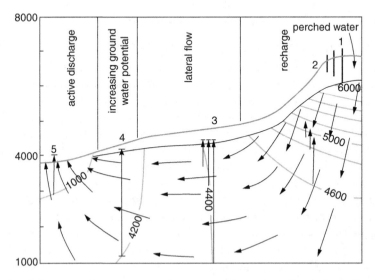

Fig. 4.2 Schematic of groundwater flow system based on observed relationships in typical great basin flow systems, Nevada, USA. (Mifflin, 1968, Fig. 5, p. 12).

Also, the water table is at a greater depth and its annual fluctuation is greater in recharge areas than in discharge areas, (Fig. 4.3a,b; Tóth, 1962a, 1984; Meyboom *et al.*, 1966; Mifflin, 1968; Freeze, 1969; Fogg and Kreitler, 1982). Although the type of these conditions is topography dependent, their magnitude, intensity and areal extent are strongly affected by the regional climate.

An important hydrologic consequence of basinal flow distribution is the regionally contrasting near-surface moisture conditions. In general, moisture contents in

Fig. 4.3 Relative simultaneous positions of the water table in different groundwater flow regimes, Andhra Pradesh, Southern India: (a) recharge area; (b) discharge area (photos by J. Tóth).

the soil and at shallow depths are deficient in recharge areas and excessive in discharge areas as compared with the throughflow regions, i.e. with regions where the
vertical water exchange at the water table is controlled solely by precipitation and
evaporation.

Depending on the climate and geology, deviations from a neutral water balance
may be manifest by imperceptible amounts of and differences in soil moisture, to
striking contrasts between parched recharge areas and marshy lowlands. Boelter
and Verry (1977) present an illustrative case from Minnesota, USA, of steady and
high annual surface runoff from a groundwater fen (discharge position) versus a
high variation and low rate of runoff from a perched bog (recharge position) of
similar areal extent (Figs. 4.4a,b).

Tóth (1984, Figs. 4.5a–c, p. 24) describes a relatively dry cashew plantation in
an upland recharge area of permeable Permian sandstone at Neyveli, Tamil Nadu,
India. It contrasts sharply with the paddy fields and flowing wells in the self-
sustained marshes at the flow system's discharge terminus, approximately 15 km
away (Figs. 4.5a–c). Moisture contrasts of this nature may be indicative of the rock's
permeability. In sand dunes, for instance, high permeability may allow significant
differences to develop in surface moisture conditions between possibly arid dune
tops and open water in inter dune depressions.

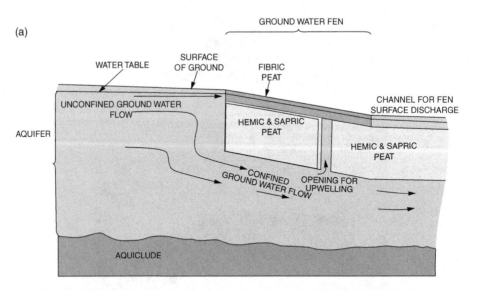

Fig. 4.4 Effect of regional groundwater flow on wetland hydrology: (a) schematic
flow pattern in groundwater fen; (b) stream-flow (surface runoff) duration curves
for a perched-bog (recharge regime) and a groundwater-fen (discharge regime)
type wetlands. Both areas are about 53 ha in size (Boelter and Verry, 1977; (a):
Fig. 7, p. 11; (b): Fig. 15, p. 16).

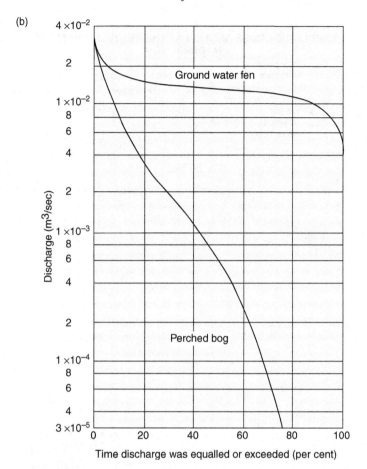

(b)

Discharge (m³/sec)

Ground water fen

Perched bog

Time discharge was equalled or exceeded (per cent)

Fig. 4.4 (cont.)

4.4.2.2 Chemistry and mineralogy

Regional groundwater flow can have many different, significant, and visible effects on the chemistry and mineralogy of the rocks and water at and beneath the land surface. Typically in the direction of flow and with increasing depth: the content of total dissolved solids (TDS) increases; the anions change from bicarbonate through sulfate to chloride and the cations from calcium–magnesium through calcium–sodium, sodium–calcium to sodium; carbon dioxide and free oxygen decrease; redox potential changes from positive to negative; and the pH from acidic to basic (Figs. 4.1, 4.6; Chebotarev, 1955; Back, 1966; Sastre Merlin, 1978; Jankowski and Jacobson, 1989; Stuyfzand, 1993; Domenico and Schwartz, 1997). Due partially to these conditions, dissolution and leaching of minerals dominate in intake areas, whereas deposition and accumulation characterize discharge

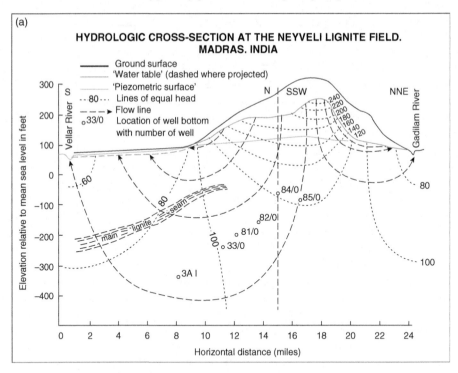

Fig. 4.5 Groundwater flow distribution and resulting contrast in surface-hydrologic conditions, Neyveli, Tamil Nadu, India: (a) groundwater flow pattern, based on data from Jones and Subramanyam (1961); (b) cashew plantation in (relatively) dry recharge area; (c) rice paddies and flowing well in discharge area (photo: Tóth).

Fig. 4.5 (cont.)

regions. Salts, such as $NaSO_4$, $NaCl$, $Ca(SO_4)_2$ and $CaCO_3$, that are brought to the land surface by discharging groundwater may be retained in the soils to cause salinization or accumulate possibly to form commercial deposits such as sodium sulfate, halite, gypsum, and calcareous tufa (Yaalon, 1963; Williams, 1970; Macumber, 1991). Weathering, dissolution, cementation and diagenesis of a wide range of rocks and minerals are natural manifestations of groundwater's chemical activity.

A well-studied example is a $NaSO_4$ deposit, 15–25 m thick, in Alberta, Canada. This deposit has accumulated from groundwater discharge in a closed topographic depression since the last glaciation, approximately 10 000 y BP (Figs. 4.7a,b and 4.8; Wallick, 1981).

4.4.2.3 Vegetation

The vegetation cover of a locality may be directly or indirectly affected, and in certain cases controlled, by groundwater flow through the effects of flow on the moisture and salinity of the area's soils (Figs. 4.1, 4.5, 4.9, 4.10). Both the type and quality of the plants are sensitive to groundwater conditions. Thus each site with different combinations of moisture and salinity (and climate) has a characteristic climax association of herbaceous and woody plants. Because most plants can tolerate a certain range of salinity and moisture, it is the association rather than the individual species that reflect the groundwater regime.

Climax associations in the relatively moisture-deficient recharge areas thrive on little soil moisture that would not sustain their discharge-area counterparts. The species of plants requiring dry, average, and wet conditions for optimal growth are called xerophytes, mesophytes and phreatophytes, respectively; salt-tolerant plants are the halophytes. Studies that show the effects of different flow regimes of

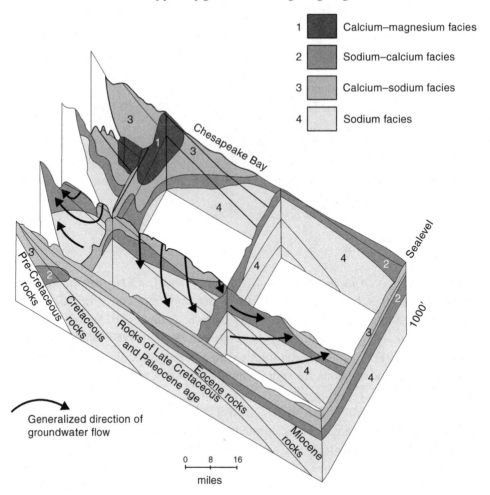

Fig. 4.6 Groundwater flow and distribution of cation facies, Atlantic coastal plain, USA (modified from Back, 1960).

groundwater on plant associations are exemplified by Leskiw (1971), Tóth (1972), Sastre Merlin (1978), Engelen and Kloosterman (1996), Batelaan *et al.* (2003). Although it does not invoke the flow-system concept explicitly, the *'Guidebook for Determining the Lithological Composition of Surface Deposits and Depth of Occurrence of Ground Waters'*, by Vereiskii and Vostokova (1966), uses a detailed, flora-style presentation of herbs, shrubs, and trees and occasionally flow arrows, to indicate relations among conditions of soil type, topography and water-table depth. The relationships are shown in such a way that, with our current understanding of regional groundwater flow, they can be interpreted in terms of hydraulic regions (recharge, discharge) of flow systems.

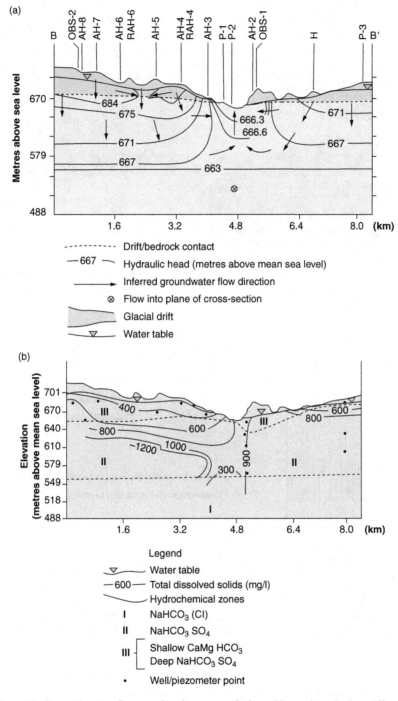

Fig. 4.7 Groundwater flow and salt accumulation, Horseshoe Lake, Alberta, Canada: (a) hydraulic cross-section; (b) distribution of chemical water types (Wallick, 1981, Figs. 3 and 4, pp. 253 and 254).

Fig. 4.8 Aerial view of groundwater-flow generated NaSO$_4$-accumulation, Horse-shoe Lake, Alberta, Canada (photo by J. Tóth).

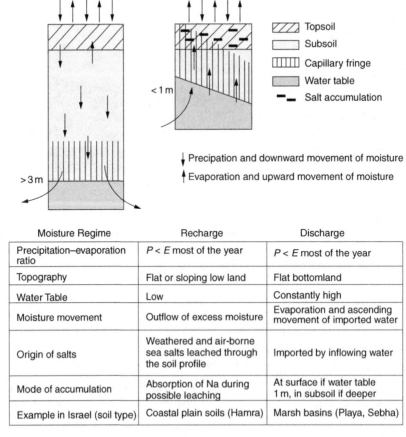

Moisture Regime	Recharge	Discharge
Precipitation–evaporation ratio	$P < E$ most of the year	$P < E$ most of the year
Topography	Flat or sloping low land	Flat bottomland
Water Table	Low	Constantly high
Moisture movement	Outflow of excess moisture	Evaporation and ascending movement of imported water
Origin of salts	Weathered and air-borne sea salts leached through the soil profile	Imported by inflowing water
Mode of accumulation	Absorption of Na during possible leaching	At surface if water table 1 m, in subsoil if deeper
Example in Israel (soil type)	Coastal plain soils (Hamra)	Marsh basins (Playa, Sebha)

Fig. 4.9 Schematic salt balance in the soil zone as function of the groundwater-flow regime (after Freeze, 1969, Fig. 3, p. 8, modified from Yaalon, 1963, Fig. 2, p. 118).

(a)

(b)

Fig. 4.10 Heavy salinization of agricultural soil, southern Alberta, Canada: (a) aerial view; (b) close up view (photos by J. Tóth).

4.4.2.4 Soil and rock mechanics

Soil and rock mechanical manifestations of groundwater's geologic agency include moisture-weakened soft soils, liquefied quick ground, mud flows and landslides. In cases where they are due to groundwater, these phenomena occur mainly in discharge areas. Here, increases in pore pressures due to ascending flow (i.e. super-hydrostatic pore pressures and superhydrostatic pore-pressure gradients) reduce the effective stresses between the soil's grains whereas relatively high water levels enhance lubrication. In combination, the two factors tend to diminish the shear strength of the soil or rock and increase their vulnerability to erosion and mass wasting. The actual form of the basically weakened soil- or rock-mechanical conditions depends largely on the hydrogeologic environment in which they occur. Soft marshy grounds, caused by ascending groundwater and often combined with

Fig. 4.11 Extensive area of liquefied and salinized soil, Southern Alberta, Canada (photo by J. Tóth).

phreatophytic vegetation and saline soils, may cover extensive areas in low-lying plains or bottoms of mountain valleys with relatively homogeneous near-surface soil/rock material (Fig. 4.11; Tóth, 1966a; Clissold, 1967; Srisuk, 1994). Liquefied conditions of limited areal extent (quick sand, 'soap holes', 'mud volcanoes') may be generated in the above situations, in which localized, relatively highly permeable lenticular heterogeneities focus upward flow through reduced cross-sections in the rock (Fig. 4.12). This phenomenon is known to occur in flood plains; valley bottoms; sea-side beaches and lake shores; desert wadis; in mines and wells; and under dams, dykes and levees (piping), in which cases they have been known to facilitate destructive floods. Significantly, their discharge water is usually fresh in mountainous regions, whereas the water can be highly saline in plains (Farvolden, 1961; Ihrig, 1966; Tóth, 1972, 1984).

In environments where slopes are dominant components of the morphology, the state of the geologic materials weakened by groundwater may be manifest by creeps, slumping, mudflows, landslides, and other forms of mass wasting. Owing to the generally high intensity, small dimensions and shallowness of the flow systems, high water salinity is not normally present in these environments, but excessive moisture is a common attribute (Zaruba and Mencl, 1969; Deere and Patton, 1971; Cherry *et al.*, 1972; Freeze and Cherry, 1979).

4.4.2.5 *Geomorphology*

The geomorphologic manifestations of groundwater's activities range from the obvious and well known to masked and unrecognized. Karst development in

Fig. 4.12 Locally liquefied ground due to near-surface sand lens, Central Alberta, Canada Canada (photo by J. Tóth).

limestone, dolomite, gypsum, and halite has long since been attributed to the action of groundwater. The effects of flow systems and their dynamics, distribution, and patterns on the evolution, depth, and geometry of karst caves, sink holes and channels, as well as on karst-water chemistry, have been extensively studied and reported on (Figs. 4.13a,b; Bedinger, 1967; LaFleur, 1984; Paloc and Back, 1992).

The genetic link between geysers and groundwater flow systems is also well understood, although less widely recognized. In discussions of the workings of geysers, attention is usually focused on the 'plumbing' systems and heat source. Although both of these attributes are indispensable components of a geyser-producing hydrogeologic environment, their principal role is to turn an otherwise regular discharge of gravity-flow systems of cold water into spectacularly functioning hot-water fountains. In general, geysers and geyser basins epitomize the many-faceted geologic effects of moving groundwater. In addition to their environmentally generated cyclic discharge of hot water, other manifestations of groundwater flow in common association with geysers are positive water balance, saline soils, mineral deposits, phreatophytic vegetation, quick ground, and other soil and/or rock-mechanical phenomena.

Groundwater-induced soil and/or rock mechanical weaknesses and, consequently, increased vulnerability to erosion often grow into major geomorphologic features. Such features include: (1) head-ward erosion started from springs or quick ground, possibly developing into gullies and stream valleys; (2) valley flanks sloping at different angles on opposite sides of streams that run parallel to the slopes' strike (consequent streams), with the surface on the higher side being concave and less steep. The asymmetry is caused by increased erosion and weakening due to the discharge of groundwater received from the higher ground (Vanden Berg, 1969);

(a)

(b)

Fig. 4.13 Karst development due to groundwater flow, northern Alberta, Canada: (a) sinkhole in recharge area; (b) gypsum cave in discharge area (photos by J. Tóth).

(3) stream-bank erosion, possibly resulting in lateral translation and concave bend-
ing of the thalweg toward the weakened area (Clissold, 1967); (4) mud flows and
landslides, induced and/or enhanced by groundwater discharge, which may leave
permanent effects on the shape of hills and mountains and may accumulate into
mounds of debris, dam streams, and otherwise modify the landscape, particularly
if occurring repeatedly.

Perhaps the least generally appreciated geomorphologic manifestations of
groundwater flow are due to its discharge in cold weather or cold climates, namely
the various forms of soil-mechanical and ice phenomena. Ice may accumulate in
the winter and in cold climates in discharge areas even where all available water
is used up by evapotranspiration in the summer time. The continual discharge of
relatively warm groundwater is often marked by soft, yellowish slush at the contact
between the ice and mineral soil (Figs. 4.14a,b). On flat terrains, the ice accumulates
as extensive fields, whereas in situations where discharge is concentrated either by
topographic features or geologic inhomogeneities, ice mounds may form.

Frost heaving and frost mounds constitute a soil-mechanical group of cold
weather/climate manifestations of groundwater flow (Holmes *et al.*, 1968; McGinnis and Jensen, 1971; Mackay, 1978). The heaving, mounding and possibly
fracturing of the ground is due to the gradual growth of frozen groundwater masses,
made possible by the continual supply of additional water at the discharge ends of
flow systems. Damage to highway pavements in Illinois and Alberta was reported

(a)

Fig. 4.14 Winter-time effects of groundwater flow on surface hydrologic conditions in boreal forest, NE Alberta, Canada: (a) relatively dry recharge area; (b)
accumulation of frozen groundwater in the discharge area (photo by J. Tóth).

(b)

Fig. 4.14 (cont.)

by Williams (1968) and Tóth (1984), respectively, and ice-mound formation was attributed to groundwater discharge in permafrost regions by Müller (1947).

In a prairie environment of Alberta, Canada, a seasonal 'frost blister', or 'frost mound', rises from the flat bottom of a spring-eroded circular depression each fall in early November. The process starts after ground frost prevents the continued discharge of groundwater through the spring's orifice and its surroundings. The mound is approximately 1.5 m high and 20 m in diameter. Throughout the winter, mounting pore pressure and a growing ice lens keep raising the ground surface until spring thaw breaks the frost seal and melts the ice lens in late March. By the end of April, the ground sinks back to its original flat and level surface, the previously open tension cracks heal, and water with increased salinity discharges through the crater-like collapsed apex of the former mound. The annual phenomenon has been observed for several decades (Figs. 4.15a–c; Tóth, 1972, 1984).

4.4.2.6 Transport and accumulation

The role of groundwater as an agent of subsurface transport and accumulation is manifest in nature by a broad range of diverse, some spectacular and some economically important, phenomena, which include: geothermal temperature patterns; sedimentary sulfide ores; roll-front and tabular uranium deposits; hydrocarbon accumulations, halos, and seeps; and eutrophication of surface-water bodies.

Moving groundwater can create systematic patterns of heat distribution in drainage basins owing, as seen earlier, to its ability to exchange heat with its ambient environment. As a result, descending cold waters reduce the temperatures and temperature gradients in recharge areas below the values which would be due solely to conductive dissipation of geothermal heat. Conversely, ascending warm waters cause positive anomalies of geothermal heat and gradients in the discharge areas (Figs. 4.1, 4.16).

(a)

(b)

Fig. 4.15 Seasonal frost mound in groundwater discharge area, Central Alberta, Canada: (a) flat surface of salt-affected frost-free ground near orifice (cribbing) of abandoned spring, October; (b) tension cracks developing in ground surface lifted by increased pore pressure due to water discharge blocked by ground frost, December; (c) final stage of raised mound before collapse, March (photo by J. Tóth).

(c)

Fig. 4.15 (cont.)

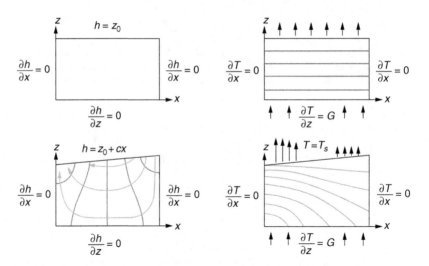

Fig. 4.16 Effect of groundwater flow on temperature distribution in the unit basin (modified from Domenico and Palciauskas, 1973, Fig. 1, p. 3805).

Flow-induced geothermal anomalies can be accentuated by increased flow rates through highly permeable fault zones and bedding planes, often leading to the development of naturally occurring thermal springs. Geothermal effects of flowing groundwater are well studied and well documented as, for instance, in Smith and Chapman (1983); Bethke (1985); Romijn *et al.* (1985); Rybach (1985); Beck *et al.* (1989); Van der Kamp and Bachu (1989); Deming *et al.* (1992); Deming (2002); and Anderson (2005) (with 194 references).

The effect of gravity-driven groundwater flow on geothermal heat distribution was convincingly demonstrated by Lazear (2006). The study is based on 3D numerical simulations of groundwater flow and heat transport constrained by field observations of levels and temperatures of well and spring water, location of springs and the type of flow systems feeding them, and measured base-flows of streams. The study area was a single watershed of approximately 600 km^2 in areal extent and nearly 1500 m relief, located on the south flank of the Grand Mesa on the western slopes of the Rocky Mountains near Cedaredge, west-central Colorado, USA. The study's objective was 'to model the steady-state dynamic equilibrium of the hydrogeological and thermal systems' (Lazear, 2006, p. 1590).

Groundwater flow and heat transport were simulated independently: the distribution of hydraulic head was computed by MODFLOW and used as input to the heat transport program executed by a custom code written by the researcher. The results were constrained by components of the basin's water balance estimated from hydrological measurements. Results of the groundwater flow modelling were displayed in two ways: a plan view map of the flow paths, and vertical cross-sections of the flow and temperature distributions. The plan view map (Lazear, 2006, Fig. 9; not reproduced here) shows a number of important and characteristic patterns of groundwater flow, namely: (i) three typical ranges of flow-path lengths: 0.5–5, 5–10, >10 km, and springs at flow-system terminations. In general, the short (and shallow) paths are associated with local topographic gradients while long flow paths penetrate to depths of several kilometres and are influenced by large-scale topographic features; (ii) shallow flow paths cross laterally over deep flow; (iii) flow paths converge toward known springs; (iv) flow paths of different lengths converge to common discharge areas where their waters may mix; (v) flow converges from a wide region into a small discharge area suggesting that proximate springs may have different water-source areas.

From the four vertical hydraulic cross-sections published by Lazear (2006, Fig. 11, profiles 1, 2, 3, 4, p. 1590) the groundwater flow pattern of 'profile 2' is reproduced here in Figures 4.17(b,c). The flow patterns of all four of Lazear's sections, exemplified by 'profile 2', confirm the features observed in the plan view, including: divergence of flow in recharge areas under topographic highs and convergence into discharge points beneath topographic lows; flow from diverse source areas may be focused and channelled into common discharge regions; mapped springs occur where flow paths converge towards the land surface. Examples of such situations are shown in Figure 4.18 (Lazear, 2006, Fig. 12) with some details of simulated flow patterns associated with mapped springs. 'Nearly all mapped springs are associated with increased upward flow at the water table in the numerical solutions' (Lazear, 2006, p. 1595).

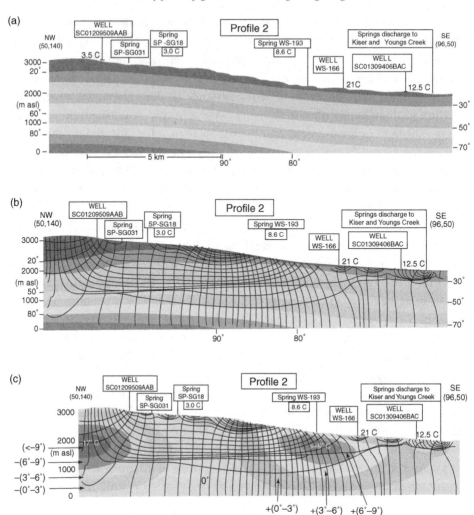

Fig. 4.17 Modelled conditions of subsurface water flow and heat transport along a northwest–southeast oriented vertical cross-section, Tongue Creek watershed, Delta County, Colorado, USA: (a) temperature distribution due solely to conductive heat transport; (b) groundwater flow and temperature distribution due to conductive and convective heat transport; (c) anomalies due to heat convection by groundwater in the conductively generated temperature field (modified from Lazear, 2006, respectively, Fig. 14, 'profile 2'; Fig. 16, 'profile 2'; and Fig. 18, 'profile 2').

Heat transport and temperature distribution through the basin were simulated for two different solutions: one for conduction alone, and one for conduction, convection, and heat from hydraulic-head loss (Figs. 4.17a and b, respectively Lazear Figs. 14 and 16 'profile 2'). The conductive solution (Fig. 4.17a) illustrates the distribution of temperature expected from the interaction of regional heat flow with

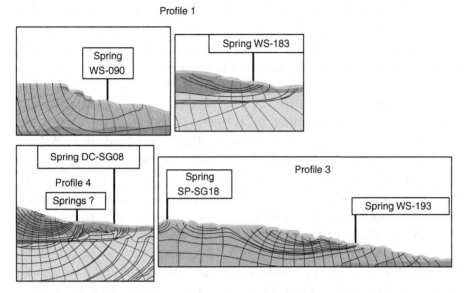

Fig. 4.18 Details of modelled flow patterns associated with mapped spring locations. Shading indicates hydrostratigraphic units (modified from Lazear, 2006, Fig. 12; profiles 1 and 4 are not reproduced in the present work).

the local topography in the absence of regional groundwater flow. The calculated temperatures were compared with observed temperatures in water wells, which resulted in a root mean squared error of 5.0°C.

The solution including convection is shown in Figure 4.17(b). The effect of groundwater flow is well illustrated by the downward shift of isotherms under recharge areas and their rise in regions of ascending flow. In this case the comparison with observed values resulted in a root mean squared error of 3.04°C, which is a nearly 2°C improvement over the hydrostatic, i.e. pure conductive case. Figure 4.17(c) shows explicitly the amount, extent, and position of the temperature changes caused by the regional groundwater flow.

Mississippi Valley-type ore deposits are considered by various researchers to be the result of the mobilization, transport and accumulation of metal ions by regional groundwater flow (Baskov, 1987; Garven *et al.*, 1993, 1999). According to these hypotheses, metal ions travel in sulfate brines that are reduced in discharge areas by sulfate-reducing bacteria, and by reaction with H_2S and/or organic matter, such as coal, peat and methane.

Theories based generally on the above processes have been advanced to explain the origin of roll-front and tabular type uranium deposits (Butler, 1970; Galloway, 1978; Sanford, 1994; Raffensperger and Garven, 1995a, 1995b). The key element in these processes is the transport of metals by groundwater from oxidizing recharge regions, where uraninite is highly soluble, to the reducing discharge environment, where it precipitates. Changing geometries of the flow systems may result in remigration of previously formed deposits (Fig. 4.19).

Certain types and many specific cases of petroleum (oil, gas) accumulations around the world have been attributed by numerous authors to the effect of basinal groundwater flow (Munn, 1909; Rich 1921; Hubbert, 1953; Hodgson, 1980; Tóth, 1980, 1988; Harrison and Summa, 1991; Verwij, 1993, 2003; Sanford, 1995).

According to the 'Hydraulic theory of petroleum migration' (Fig. 4.20; Tóth, 1980), hydrocarbons can be mobilized in mature source rocks or carrier beds by groundwater flow and drawn by its systems toward discharge areas in any of several different forms such as bubbles, globules, stringers, ionic solution, emulsion and micelles. En route to a region of flow convergence, the concentration of hydrocarbons increases. Entrapment, thus local accumulation, can be caused by minimum fluid potentials for petroleum, mechanical screening by pore-size reduction at strata boundaries, and by capillary pressure barriers (Hubbert, 1953; Tóth, 1988). The effectiveness of the latter two mechanisms can be augmented by growth of the hydrocarbon particles due to coagulation during transportation. The growth and aggregation of particles are enhanced, in turn, by the decreasing solubility of petroleum in the rapidly decreasing temperatures and pressures in the discharge areas' ascending waters. Regional discharge areas offer, therefore, ideal conditions for accumulation and entrapment and are, indeed, the sites of many large petroleum deposits (Fig. 4.20; Bars *et al.*, 1961; Hitchon and Hays, 1971; Chiarelli, 1973; Tóth, 1980; Wells 1988; Deming *et al.*, 1992).

Hydrocarbon particles that escape entrapment may continue their migration towards the surface and may form concentration anomalies in groundwater, soil gas, plant tissue, sea-bottom sediments, saturated pools as halos, seeps or springs. During their migration, the hydrocarbons cause chemical reduction in the traversed column of earth material. The affected subsurface environment is called the 'geochemical chimney', and it is used extensively in geochemical exploration for petroleum (Schumacher and Abrams, 1996). Tóth (1996) has attibuted the less-than-expected effectiveness of geochemical exploration to a lack of understanding of regional groundwater flow among the exploration community.

One possible cause of eutrophication of surface-water bodies is the transport of plant nutrients by groundwater into lakes. Shaw *et al.* (1990) show that phosphorus and nitrogen, derived from fertilizers and lake-side cottage cesspools, are transported in systems of groundwater flow and discharged over large portions of lake bottoms, causing anomalously rich growths of aquatic plants. Eutrophication, in such cases, is a manifestation of groundwater's geologic agency.

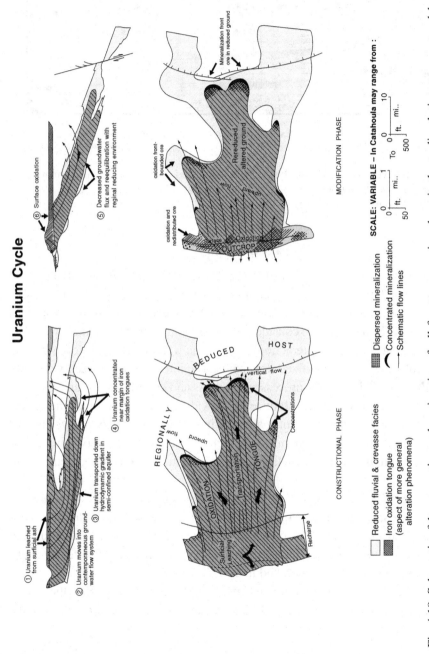

Fig. 4.19 Schematic of the generation and remigration of roll-front type uranium deposits due to dissolution, transport and deposition by gravity-drive groundwater flow (modified from Galloway, 1978, Fig. 10, p. 1673). The two principal phases of the uranium cycle interpreted for Catahoula fluvial systems.

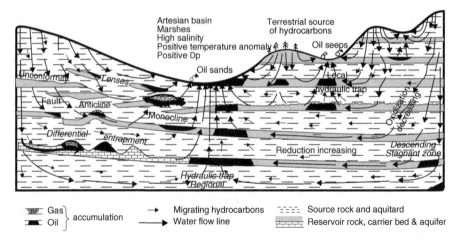

Fig. 4.20 Hydraulic and hydrodynamically enhanced geologic traps in regionally unconfined fields of gravity flow of formation waters (Tóth, 1988, Fig. 10, p. 491, modified from Tóth, 1980, Fig. 44, p. 163).

4.5 Summary

The intent of Chapter 4 is to focus attention on the role of groundwater as a general geologic agent. It seeks to provide a summary overview of those types and instances of natural processes and phenomena that have been recognized as manifestations of that agency. Two causes are considered to be fundamental to groundwater's role as a geologic agent: *in-situ* interaction between the water and its environment, and hierarchically nested systematized flow paths. The interactions between water and environment result in various processes, products and conditions. In turn, the hierarchically nested flow systems produce basin-wide self organization of the effects of interaction on different scales of space and time. Ten chemical, two physical, and three kinetic processes are recognized through which interaction between groundwater and its environment, as well as organized distribution of effects, may take place.

(1) Chemical processes: dissolution, hydration, hydrolysis, oxidation-reduction, attack by acids, chemical precipitation, base exchange, sulfate reduction, concentration, and ultrafiltration or osmosis. (2) Physical processes: lubrication and pore-pressure modification. (3) Kinetic processes: the transport of water by itself, aqueous and non-aqueous matter, and heat.

Owing to the transporting ability and spatial patterns of basinal flow, the effects of interaction are cumulative and distributed in harmony with the geometries of the flow systems.

These processes are manifest in a great variety of natural phenomena. Although the number and diversity of the manifestations are great, they can be considered as environmentally modified versions of six basic types: (1) hydrology and hydraulics; (2) chemistry and mineralogy; (3) vegetation; (4) soil and rock mechanics; (5) geomorphology; (6) transport and accumulation.

In essence, the general effects of groundwater's geologic agency are: phenomena of depletion and accretion of water, metallic and nonmetallic solids, hydrocarbons, and heat; soil and rock mechanical instabilities, possibly developing into geomorphologic features; diagenetic mineral changes; changes in vegetation cover; eutrophication of surface-water bodies; and systematic areal distribution, or self-organization.

5

Practical applications: case studies and histories

Properties, controlling factors, and natural effects of gravity-driven groundwater flow were discussed in the previous chapters. Based on that understanding including theory, measured fluid-dynamic parameters, and field manifestations, it is possible to infer spatial patterns of flow systems that, in turn, can be useful in solving utilitarian problems of a widely variable nature. Examples include, but are far from limited to: basin-scale assessment of groundwater resources; locating favourable spots for water-supply wells; searching for areas suitable for specified land uses (the required characteristics of the area are specified, e.g., 'find an area suitable for a municipal sewage lagoon') or, the other way around, developing a site for use that is best suited to its natural characteristics (the characteristics of the available site are known, e.g., 'this site is suitable for a municipal sewage lagoon'); predicting the consequences entailing modifications to current groundwater-related conditions; amelioration of undesirable environmental conditions (e.g., favourable or unfavourable ecological changes); diagnosing the causes of seemingly anomalous conditions in: pore pressures, soil mechanical behaviour (liquefaction, landslides), soil- and groundwater-chemistry, vegetation and geothermal heat flow, just to name a few. Indeed, the recognition and solution of such problems are limited only by the depth of understanding, scope of experience, and imagination of the hydrogeologist involved. The practical use and usefulness of groundwater flow-system studies is perhaps best epitomized by the various schematized versions of the composite flow-system pattern adorning several major texts, monographs and posters (Fig. 5.1).

The case studies and histories presented in Chapter 5 are divided into five arbitrary and, necessarily overlapping, general topics: (1) the characterization and portrayal of regional hydrogeologic conditions; (2) problems related to recharge- and discharge-regimes of gravity-flow systems; (3) site selection for nuclear-fuel waste repositories; (4) utilization of fluid-dynamic parameter anomalies; and (5) exploration for petroleum and metallic minerals. The examples are not intended to

Fig. 5.1 Schematized version of the composite flow pattern as logo: Freeze & Cherry (1979); Instituto de Geofízico (1985); Pollock (1989); Shibasaki and Research Group (1995).

cover complete cases of problems, situations or solutions. They are meant rather to illustrate only those specific aspects of the selected questions the solution of which depends chiefly on the knowledge and/or analysis of gravity-driven flow systems.

5.1 Characterization and portrayal of regional hydrogeologic conditions

The type, size, and distribution of gravity-driven groundwater flow-systems and their environmental consequences are intrinsic components of a region's hydrogeology. Such information can be shown on maps and cross-sections for practical purposes, and the practice has been widely adopted since the introduction of the flow-system concept in the early 1960s. Some international examples are presented below illustrating different methods, objectives and scales for depicting groundwater flow systems on hydrogeological maps and sections.

5.1.1 The hydrogeological reconnaissance maps of alberta, canada

Perhaps the first hydrogeological maps that portrayed groundwater flow-systems and their known environmental effects as standard components of the maps' contents were produced in the programme of 'the hydrogeological reconnaissance mapping of Alberta' (Tóth, 1977, p. 1). The programme was initiated in 1968, lasted for 12 years, covered the entire 660 000 km^2 area of the province of

Alberta, and produced 48 sheets of eight-coloured hydrogeological maps at scales of $M = 1 : 125\,000\ 250\,000$ and $500\,000$, depending on available data.

Figure 5.2 shows parts of the hydrogeological map and a hydrogeological cross-section of the Grimshaw–Chinook Valley area produced as the mapping programme's prototypes (Tokarsky, 1971; Tóth, 1977). Lateral flow directions of groundwater are shown on the map by hydraulic-head contours, as usual. However, where known or inferred, the sense of the vertical flow direction also is shown. Small crosses on the contours indicate downward flow and beads mean upward flow on the map, and arrows point in the flow direction on the section.

Hydraulic-heads were contoured in the vertical also on some cross-sections produced in later years. Regions of flowing wells, a diagnostic sign of groundwater discharge, are circumscribed by a stipple band, while isolated artesian wells are shown by upward pointing arrows. Springs, saline soils, unstable soil mechanical conditions are indicated by a coherent set of standardized map symbols (Badry, 1972).

The philosophy of the maps' design was based largely on the conclusion that knowledge of groundwater flow-systems affords a diversity of uses of the maps. The

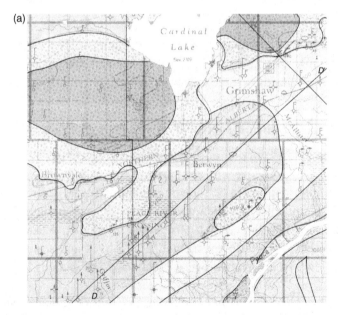

Fig. 5.2 Hydrogeological reconnaissance map of the Grimshaw–Chinook Valley area, Alberta, Canada: (a) detail of the main map (shading indicates potential total production rates of single wells to depth of 305 m (1000 ft); (b) detail of section $D-D'$ (shading indicates potential production of single wells in the hydrostratigraphic unit) (from Tokarsky, 1971).

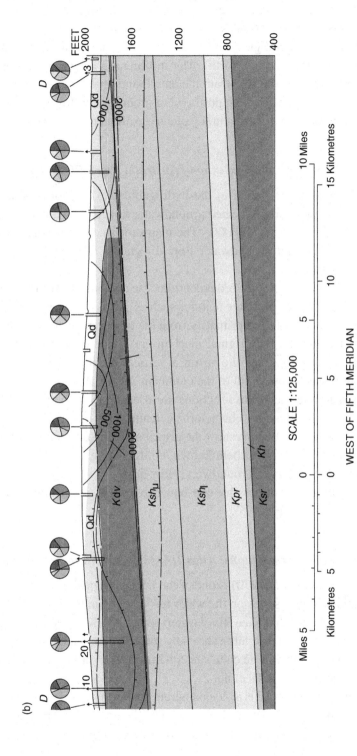

Fig. 5.2 (cont.)

hydrogeological conditions were presented, therefore, in terms of basic parameters. 'Thus the majority of information is shown as components of the groundwater regime and of the hydrogeological environment rather than in terms of specific uses (domestic, industrial, and so forth). The information sought by a map user will normally have to be composed from the data found in the maps' (Tóth, 1977, p. 6). One of these data is the explicit portrayal of groundwater flow systems, even though often limited in detail and accuracy due to insufficient basic data.

5.1.2 *The hydrogeological map of Australia (M=1:5 000 000)*

Some of the main items presented in the hydrogeological map of Australia are the explicitly and/or indirectly indicated groundwater flow patterns. Published at a continental scale of $M = 1 : 5\ 000\ 000$, 'The map shows aquifer types, lithology, salinity, potentiometry, and groundwater flow directions, and major abstraction sites' (Lau *et al.*, 1987, p. vi).

Groundwater flow lines, generalized potentiometric contours, proven groundwater discharge areas, specific discharge features such as playas and groups of springs, and groundwater divides inferred primarily from the topography, are depicted in a distinctive violet colour on the original map (in grey scale on Fig. 5.3a). Major groundwater flow directions are shown on a vertical cross-section also, which traverses the entire east–west width of the continent (Fig. 5.3b).

Relatively large-scale local cross-sections are used in an accompanying short booklet (21 pages), to illustrate characteristic features of the major hydrogeologic provinces. The cross-sections show the principal hydrostratigraphic units and groundwater flow directions or, where known, entire flow systems. The origin of playas is attributed explicitly to groundwater (Fig. 5.4).

A specific diagram warns of possible consequences of soil salinization as a direct result of human activities (Fig. 5.5).

5.1.3 *'Hydrologic Investigations: Atlas HA-339', NW Minnesota; USGS*

'*Atlas HA-339*' (Winter *et al.*, 1970) is one of the US Geological Survey's 'Hydrologic Investigations' report series on the water resources of individual watersheds. The report deals with the Wild Rice River watershed of NW Minnesota. It presents the intended information on four large sheets by means of annotated maps, graphs, sections and drawings of the basic data, site conditions, and summary conclusions. Groundwater is the subject of sheet no. 2.

The centrepiece of sheet no. 2 is a block diagram cut into a north- and a south-half along an east–west striking section (Fig. 5.6). The diagram shows the surface of the watershed divided along a N–S line into two different regimes of groundwater

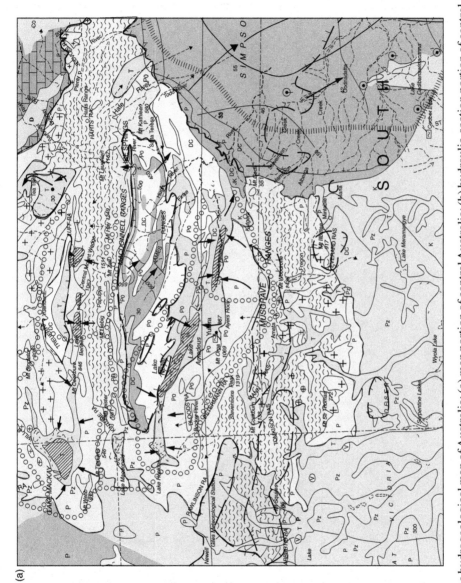

(a)

Fig. 5.3 The hydrogeological map of Australia: (a) map portion of central Australia; (b) hydraulic-section portion of central Australia (modified from Lau et al., 1987).

(b)

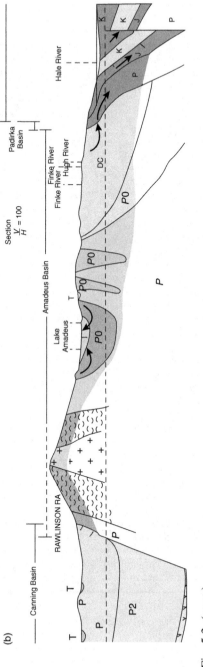

Fig. 5.3 (cont.)

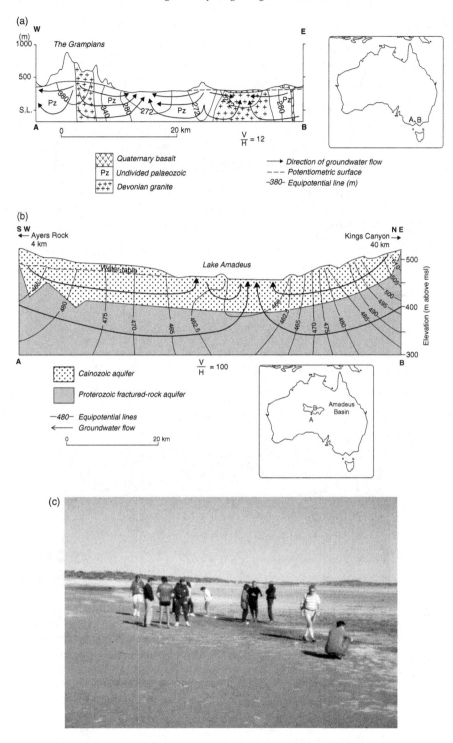

Fig. 5.4 Groundwater flow patterns in selected sub-basins: (a) Ballarat, Victoria; (b) Amadeus Basin, Northern Territory; (c) extensive salt playa in Lake Amadeus (a) and (b) Lau *et al.*, 1987, Fig. 18, p. 18 and Fig. 5, p. 7, respectively; (c) photo by J. Tóth.

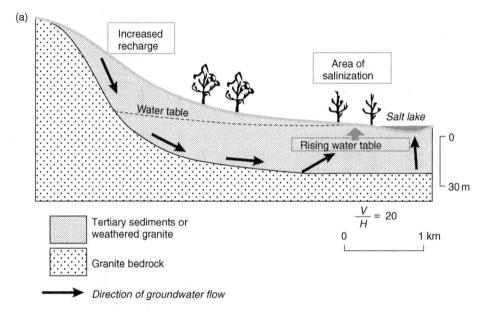

Tertiary sediments or weathered granite

Granite bedrock

Direction of groundwater flow

Fig. 5.5 Artificially induced flow-system modifications and consequent environmental effects in discharge areas: (a) land clearing for agriculture in southwest Australia: reduced transpiration, increased recharge, rising water table and salinization (Lau *et al.*, 1987, Fig. 17, p. 17); (b) damming the outlet of nearby lake to improve recreational potential, Murray River basin, southeast Australia: rising water table (increased discharge), salinization, water logging and death of eucalypt forest (photo by J. Tóth).

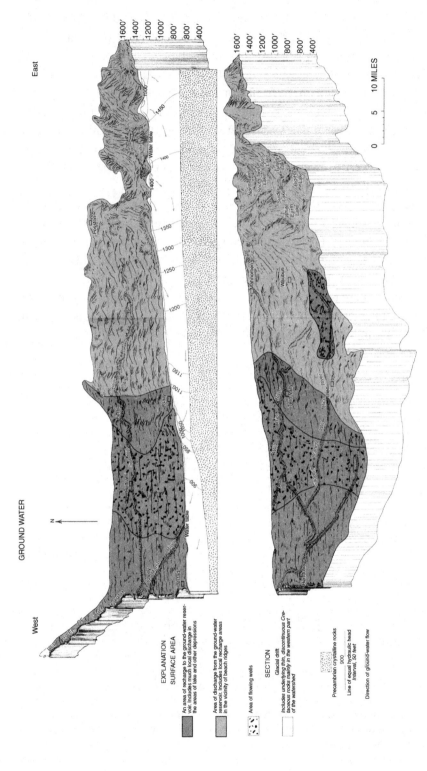

WATER ENTERS THE GROUND-WATER SYSTEM LARGELY IN THE ROLLING UPLANDS OF THE MORAINAL AREA.

Fig. 5.6 Distribution of groundwater recharge- and discharge- areas, Wild Rice River watershed, NW Minnesota, USA. Black spots in discharge area represent flowing wells (Winter *et al.*, 1970, sheet 2 of 4).

hydraulics. The eastern half is a recharge area. It is hilly, covered by glacial drift, and rises to elevations between 1200 and over 2000 ft (366 and 610 m) above sea level. The western half is a slightly less extensive discharge region of flat lake-plain deposits ranging between 800 and 1200 ft (244 and 366 m) in elevation.

A vertical hydraulic section on the front panel of the block's northern half shows entire local flow systems, and some recharge portions of large-scale flow systems, on the watershed's eastern side. The systems are induced by the area's hilly morainal topography and, respectively, by its regionally high altitude. The lower lying lake-plain region on the west is a general area of groundwater discharge, as indicated by extensive areas of flowing wells.

A genetic relation between the chemical composition and flow pattern of groundwater is demonstrated on a separate, three-pane fence diagram of sheet no. 2 (Fig. 5.7). The Chebotarev (1955) sequence (generally characteristic of changes in chemical composition of gravity-driven flow systems, although he did not mean or recognize it that way!) appears to develop in the area. In local (short) flow systems, and in recharge areas of longer flow systems, the main chemical type of the water is calcium–magnesium bicarbonate. It changes through calcium–magnesium sulfate and/or sodium bicarbonate, to sodium chloride in the flow direction of the longer systems.

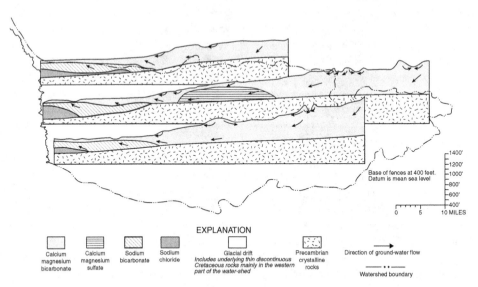

EXPLANATION

Calcium magnesium bicarbonate	Calcium magnesium sulfate	Sodium bicarbonate	Sodium chloride

Glacial drift
Includes underlying thin discontinuous Cretaceous rocks mainly in the western part of the water-shed

Precambrian crystalline rocks

→ Direction of ground-water flow

—··— Watershed boundary

Base of fences at 400 feet. Datum is mean sea level

1400'
1200'
1000'
800'
600'
400'

0   5   10 MILES

CALCIUM MAGNESIUM BICARBONATE THE MOST COMMON TYPE OF GROUND WATER IN THE AREA, OCCURS MAINLY IN THE RECHARGE AREA

Fig. 5.7 Fence diagram showing relations between chemical composition and flow systems of groundwater, Wild Rice River watershed, NW Minnesota, USA (Winter *et al.*, 1970, sheet 2 of 4).

5.1.4 Protection and restoration of wetland ecosystems based on groundwater flow-system analysis, the Netherlands

Engelen and Kloosterman (1996) published *Hydrological Systems Analysis: Methods and Applications* devoted specifically and entirely to utilitarian applications of the characterization and analysis of gravity-driven groundwater flow systems. Part 1 of the book treats the basic principles, study methods, and possible utilizations of the flow-system concept in six chapters, while Part 2 presents thirteen case studies in as many chapters, most of them from the Netherlands.

Chapter 12 outlines a project the purpose of which was 'to provide a scientific basis for integrated water management of the (Twente) region; in particular with regard to nature management' (Engelen and Kloosterman, 1996, p. 83). Thus the study was aimed at determining the effects of human activities, such as municipal water withdrawal and modification of the surface-water network by drainage, on the natural flow systems of the past. To this end, 'The quantitative effects of those human interventions on the spatial arrangement, form and size of the natural and man-made flow systems, their fluxes and patterns of flow lines, residence times and isohypses were studied for natural and disturbed conditions in both winter and summer conditions.' (Engelen and Kloosterman, 1996, p. 83).

The study was based on groundwater-related field-data and modelling of groundwater flow covering an area of 16.5 km × 24 km. The results were presented on a series of maps and cross-sections which included: flow-line maps; travel-time maps; hydrological system maps (Fig. 5.8a,b); maps of historical and contemporary land use and wetland patterns; maps of impact on groundwater levels; vertical sections and schematic reconstructions of ecosystems and groundwater flow-systems for 1850 and 1990 (Fig. 5.9), showing the ecological effects of the modifications to groundwater flow.

Figure 5.8 shows the horizontal trajectories of the three-dimensionally calculated flow lines: for the 'original', or undisturbed, situation during the nineteenth century (Fig. 5.8a); and for the end of 1990, including artificial flow systems induced by contemporary pumping stations, but without the effects of agricultural drainage (Fig. 5.8b).

Figure 5.9 is a schematic representation of groundwater flow-patterns and the associated ecologic landscape along a general east–west vertical section in a historic (nineteenth century; Fig. 5.9a) and a modern (late twentieth century; Fig. 5.9b) situation.

The areal pattern of the contemporary distribution of the 'infiltration and exfiltration zones' was used as the basis of a proposal for the protection and reclamation of wetlands in a network of to-be-designed nature areas: 'Such a network would counteract the (existing) fragmentation of wetland ecosystems. Moreover it could

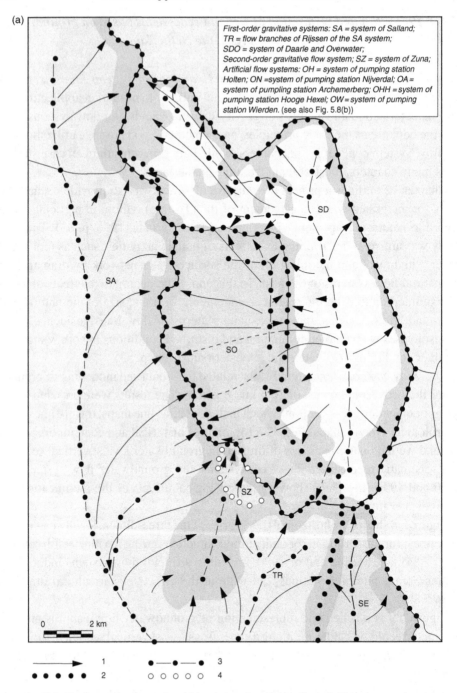

(a)

First-order gravitative systems: SA = system of Salland; TR = flow branches of Rijssen of the SA system; SDO = system of Daarle and Overwater; Second-order gravitative flow system; SZ = system of Zuna; Artificial flow systems: OH = system of pumping station Holten; ON = system of pumping station Nijverdal; OA = system of pumpling station Archemerberg; OHH = system of pumping station Hooge Hexel; OW = system of pumping station Wierden. (see also Fig. 5.8(b))

Fig. 5.8 Horizontal trajectories of the three-dimensionally calculated groundwater flow lines, Twente region, the Netherlands: (a) 'original' or undisturbed situation during the nineteenth century; (b) in 1990, including artificial flow systems induced by contemporary pumping stations, but without the effects of agricultural drainage (Engelen and Kloosterman, 1996, Plates 15 and 14, respectively).

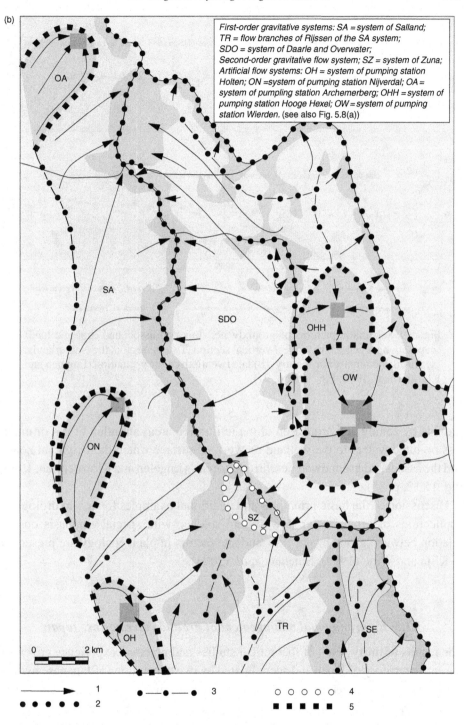

(b)

First-order gravitative systems: SA = system of Salland;
TR = flow branches of Rijssen of the SA system;
SDO = system of Daarle and Overwater;
Second-order gravitative flow system; SZ = system of Zuna;
Artificial flow systems: OH = system of pumping station
Holten; ON =system of pumping station Nijverdal; OA =
system of pumpling station Archemerberg; OHH =system of
pumping station Hooge Hexel; OW =system of pumping
station Wierden. (see also Fig. 5.8(a))

Fig. 5.8 (cont.)

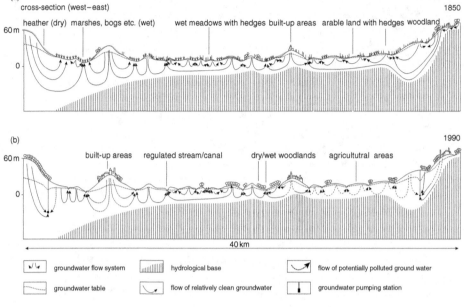

Fig. 5.9 Schematic patterns of groundwater flow and associated ecologic land-scapes along a general east–west vertical section, Twente region, the Netherlands: (a) during the nineteenth century; (b) late twentieth century situation (Engelen and Kloosterman, 1996, Fig. 12.8, p. 94).

provide by zoning and protection of the infiltration areas an influx of un- or much less polluted waters to the wetland network of surface water, their riparian zones and the associated groundwater exfiltration areas' (Engelen and Kloosterman, 1996, Fig. 11.12, p. 85).

Discussion of the basic principles and additional examples for eco-hydrological applications of groundwater flow-system studies, with special emphasis on the relation between groundwater flow and site factors in plant ecology are presented in Klijn and Witte (1999), Batelaan *et al.* (2003).

5.1.5 Environmental management of groundwater basins, Japan

The results of thirty years of theoretical studies and practical experience covering the entire territory of Japan and accumulated by two generations of Japanese hydro-geologists and engineers have been summarized in *Environmental Management of Groundwater Basins* by Shibasaki and Research Group (1995; Fig. 5.1). The authors discuss the objectives, hydrogeological principles, and methods of investigation of basin management.

Prior to the 1960s, development of groundwater resources in Japan was based on the evaluation and production of individual wells rather than on comprehensive regional evaluations (not unlike elsewhere). Notwithstanding the various laws controlling the siting and operation of water wells, problems of land subsidence, water-level declines, contamination, and sea water intrusion have made groundwater gradually an uneconomical source of supply. In the early 1960s a shift from the management of individual wells to the management of groundwater basins took place which 'was not compelled by law but was motivated by the desire to look at the groundwater movement in wider regions with the groundwater basin as the basic unit' (Shibasaki and Research Group, 1995, p. 36).

A prerequisite for successful groundwater basin management is, according to the book's authors, 'to have sufficient knowledge on the natural conditions of groundwater basins, particularly on (1) the structure of the groundwater basin which is the container of groundwater, (2) hydrostratigraphic unit which is the basic unit for groundwater flow and its structure, and (3) groundwater flow and its characteristics' (Shibasaki and Research Group, 1995, p. 45).

Gravity is considered to be the principal driving force of basinal groundwater flow in Japan, because 'the correlation between a groundwater flow system and topography is conspicuous' and 'generally the groundwater table configuration reflects the topography; and the groundwater flow system can, to some extent, be inferred from topographical structures' (Shibasaki and Research Group, 1995, p. 58–59). Starting, therefore, from the above mentioned empirical observations, the fundamental working model adopted for environmental basin management in Japan is the 'hierarchically nested flow systems', discussed in Chapter 3.

Figure 5.10 shows the conceptual flow pattern applied to the sedimentary basins of the country, with the three different orders of flow systems, namely, local, intermediate, and regional. Figure 5.11 is an example of the map-presentation of the same flow-system types, requiring significantly different spatial scales, i.e. different levels of resolution, for appropriate portrayal.

5.2 Effects of recharge–discharge area characteristics on groundwater-related practical problems

The success or failure of numerous groundwater-related practical projects depends on the recognition, exploitation, and/or control of the specific hydraulic and hydrologic characteristics of the recharge- and discharge-regimes of gravity-driven flow systems. The most salient of these characteristics are summarized in Table 5.1 (see also Figs. 2.3 and 4.1).

The probability of success of a groundwater dependent project can be expected to increase with an increase in the degree of compatibility between

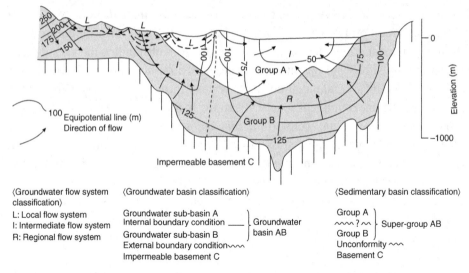

Fig. 5.10 Conceptual groundwater flow model applied to management of sedimentary basins in Japan (Shibasaki and Research Group, 1995, Fig. 3.1, p. 57).

the project's requirements and the relevant hydrogeologic characteristics of the area. Indeed, such characteristics may be dominant concerning the suitability of a site for a planned operation or facility. The cases below present the essential aspects of some projects and/or situations in which the nature of the hydraulic regime (recharge and/or discharge) has played, or can be expected to play, a decisive role in determining the suitability of a site or area for utilitarian purposes.

5.2.1 Location and development of a municipal groundwater supply, Olds, Alberta, Canada

The task of finding and developing a municipal groundwater supply for the central Alberta town of Olds in 1964 may have arguably been the first case in which the theory of gravity-driven flow was used explicitly as the basis of an approach. Since its first public-supply wells were drilled and installed twenty years earlier, in 1946, the rural town of 3000 people had been facing a chronic crisis of water shortage. The yields of countless newly drilled wells were marginal and would soon dwindle after production began, necessitating continued attempts to find new sources. Ultimately, in 1963 the town turned for help to the Research Council of Alberta, the research arm of the provincial government, and I was charged with the project.

Regional flow system under regional scale landform   Intermediate flow system under intermediate scale landform

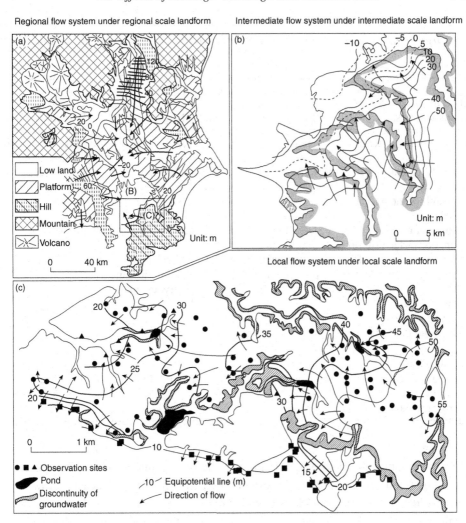

Fig. 5.11 Classification and presentation of groundwater flow systems generated by landforms of different spatial scales (Shibasaki and Research Group, 1995, Fig. 3.2, p. 59).

Two basic considerations guided my exploration strategy: one geological and one of subsurface hydraulics. Two geologic formations were known potentially to offer a prospect for a source of water of acceptable quantity and quality, namely:

(a) the unconsolidated surficial sediments of glacial drift and alluvial sands and gravels in buried bedrock valleys;

(b) the 'subaerial floodplain and lake deposits, with randomly arranged lenses, and discontinuous beds of sandstones, siltstones and claystones' of Upper Cretaceous–Paleocene age (Tóth, 1966b, p. 28).

Table 5.1. *Summary of the salient hydraulic and hydrologic characteristics of the recharge-regimes and discharge-regimes of gravity-driven groundwater flow systems*

Hydraulic/hydrologic characteristics	Flow-system Regime		Examples
	Recharge	Discharge	
Change in water-level with depth	Decrease	Increase	Figs. 4.2; 4.5, 5.56
Annual fluctuation of the water table (width of the phreatic belt)	Relatively large	Relatively small	Figs. 4.3a,b, 5.28a,b
Volume of average annual surface runoff	Relatively small	Relatively large	Fig. 4.4 a,b
Temporal variation of surface-runoff rates	Highly dependent on precipitation and relatively large	Less dependent on precipitation and relatively small	Fig. 4.4 a,b
Ratio of actual (dynamic) to nominal (static) pore pressure; P_{dyn}/P_{stat}:	< 1	> 1	Fig. 2.2
Degree and main type of groundwater mineralization	Relatively low; Ca, Mg	Relatively high; Na	Figs. 4.6, 5.7, 5.65, 5.77b

Thus the essential hydrogeologic feature of the rock framework was the absence of any distinguishable aquifer. Experience had shown that the occasional highly permeable local rock pod could not be relied upon to have sufficiently large volumes for the required sustainable rates and amounts of water production. In other words, the geological information yielded neither guidance nor limitations as to preferred areas for test drilling.

My hydraulic argument was based on the area's topographic configuration and the presence of some flowing wells. The land surface declines at an even regional rate of approximately 2 m/km from the town of Olds, located on a north–south striking topographic ridge at an elevation of 1044 m a.m.s.l. (above mean sea level) on the west, to an average elevation of 980 m at the Lonepine creek on the east; some flowing farm wells were known to occur on the west side of the creek. A regional flow system was thus postulated with its recharge area on and east of the Olds topographic ridge, and the discharge area east of the approximate midline between the ridge's crest and the Lonepine creek (Fig. 5.12; Tóth, 1966b, Fig. 16, p. 40).

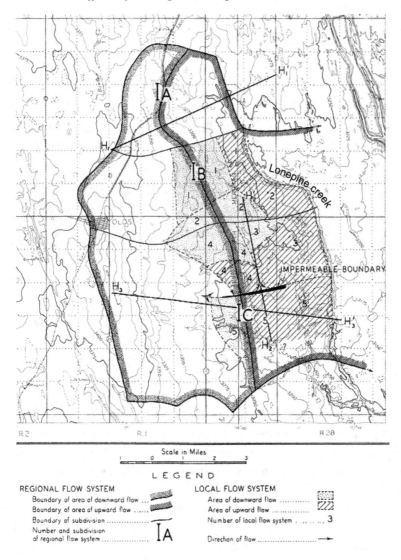

Fig. 5.12 Distribution of areas of downward and upward flow, Olds groundwater
exploration area, Alberta, Canada (Tóth, 1966b, Fig. 16, p. 40).

The decision to start test drilling in the postulated discharge area, i.e. east of the
theoretical hydraulic midline, despite the unfavorably large distance from town,
was based on the dual argument that: (a) no hydrostratigraphically preferential area
could be identified anywhere due to the rock framework's lithologic heterogeneity,
and (b) available drawdowns (a factor to which well yields are directly proportional)
can be increased on purpose by moving downslope from the hydraulic midline.
The theoretically established fact was thus invoked, namely, that hydraulic heads
increase with depth in areas of ascending groundwater flow (Section 2.1).

Fig. 5.13 'Olds Well no. 189', flowing in theoretically anticipated regional discharge area, Olds groundwater exploration area, Alberta, Canada (Tóth, 1966b, Fig. 15; photo by J. Tóth).

Results of the first three test wells supported the theoretical expectations and opened the way to a fully-fledged test-drilling (16 more wells) and pump-testing programme (Fig. 5.13).

Use of the cable-tool (percussion) drilling method facilitated accurate and multiple water-level measurements at different depths of individual wells and thus the quantitative portrayal of flow patterns. The map in Figure 5.12 shows three regional flow systems, IA, IB and IC, stretching from the topographic ridge at Olds to Lonepine Creek, with the north–south oriented hydraulic mid-line dividing the areas of regional discharge and recharge approximately half way. Local systems 1–5, with their respective recharge and discharge areas are superimposed on the regional systems (Figs. 5.12, 5.14, 5.15).

The practical result of the programme was the development of a field of three production wells. The water was pumped to the town uphill by a distance of approximately 60 m and 9 km.

5.2.2 Underestimated rates for required dewatering, and land subsidence at lignite mine in discharge area, Neyveli, Tamil Nadu, India

The Neyvely lignite field covers an area of over 200 km^2 in the coastal belt of eastern Tamil Nadu (formerly known as Madras State), India. It is located in the discharge area of a 16–18 km long regional flow system (Fig. 4.5a). Commercial interest in the possible exploitation of the estimated 230 million tons of lignite

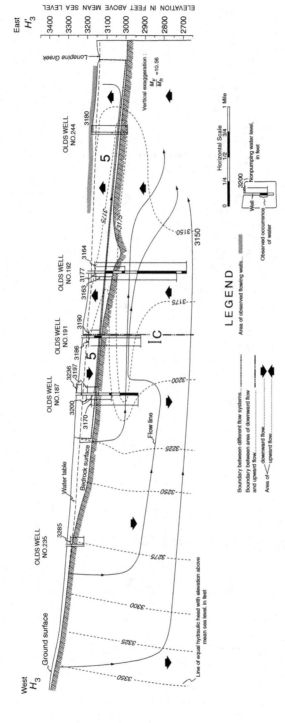

Fig. 5.14 Hydraulic cross-section $H_3 - H'_3$ (E–W), Olds groundwater exploration area, Alberta, Canada (Tóth, 1966b, Fig. 15).

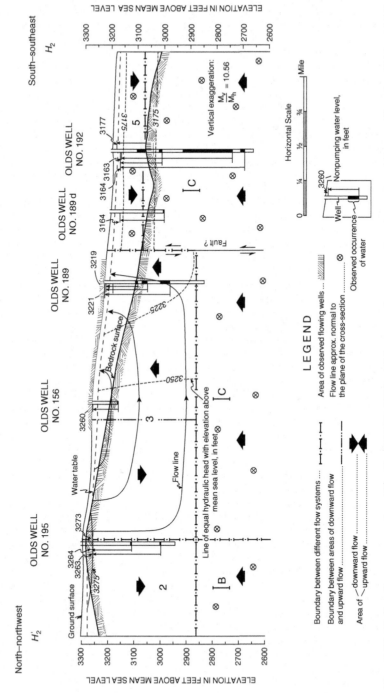

Fig. 5.15 Hydraulic cross-section $H_2' - H_2$ (NNW–SSE), Olds groundwater exploration area, Alberta, Canada (Tóth, 1966b, Fig. 15).

reserves prompted thorough geological and groundwater investigations in the early 1950s. The studies included regional-flow characterization and extensive pumping tests to determine the feasibility of pit-dewatering to facilitate mining. In a detailed technical paper Jones and Subramanyan (1961) present a summary on regional flow: 'The most significant features of ground-water movement in the Neyveli ground-water reservoir [combination of two or three thick sand and gravel beds underlying the main lignite seam; author's comment] are: regional hydraulic continuity; gentle hydraulic gradients; divergent flow from the groundwater divide about 6 miles [9.65 km] north of the pilot quarry; and converging flow south of Neyveli, in the Vellar River lowland, as a consequence of large, continuous withdrawal from flowing artesian wells' (Jones and Subramanyan, 1961, p. 289–290).

'Regional hydraulic continuity', or cross-formational communication between different stratigraphic levels, i.e. the three dimensionality of the flow field, as well as the possible outward migration of the groundwater divide in response to pumping, were thus explicitly recognized. Nevertheless, the rocks' hydraulic properties (i.e. transmissivity, T, and storativity, S) were calculated 'using the Theis non-equilibrium formula' (Jones and Subramanyan, 1961, p. 292). The method is valid, however, only for two-dimensional radial flow to wells in ideally confined horizontal aquifers. The present author's personal communication with local mine hydrogeologists in 1970 confirmed that those calculations underestimated the rates and amounts of water withdrawal required to keep the mine sufficiently dewatered. In addition, incessant abstraction had reduced pore pressures in the naturally 'overpressured' artesian aquifers so as to cause widespread land subsidence of considerable magnitude (Fig. 5.16).

5.2.3 Failure of a municipal sewage lagoon built on local recharge area, Brooks, Alberta, Canada

Ignoring the recharge character of a site, aggravated by misjudgment of the rock framework's hydraulic properties, has led to the irreparable leakage and need for decommissioning and reconstruction of the municipal sewage lagoon of a town of 8000 in southern Alberta, Canada (Tóth, 2003). Untreated sewage leaked beneath the lagoon's dykes and surfaced beyond its boundaries. It threatened also to contaminate a nearby irrigation canal used in emergencies as a source of drinking water for a hamlet downstream.

Figure 5.17 shows a mound that is the recharge area of a local groundwater flow system. Construction equipment and the lagoon's dykes are seen on top of the rise. The mound has a relative elevation of slightly over 15 m (50 feet) above the irrigation ditch that skirts the lagoon within 900 m (Fig. 5.18).

Four lithologically different rock units can be identified at the lagoon site to depths of approx. 25 m (82 ft). Owing to their hydraulically also different properties,

(a)

(b)

Fig. 5.16 Dewatering of lignite mine in groundwater discharge area, Neyveli, Tamil Nadu, S. India: (a) heavy discharge from dewatering wells surrounding open-cast pit; (b) land subsidence marked by height of top of well-casing grout above land surface, caused by underestimated rates of required pumping for mine dewatering (photos by J. Tóth).

Fig. 5.17 Recharge-area position of lagoon on local hilltop, with irrigation canal in foreground (photo by J. Tóth).

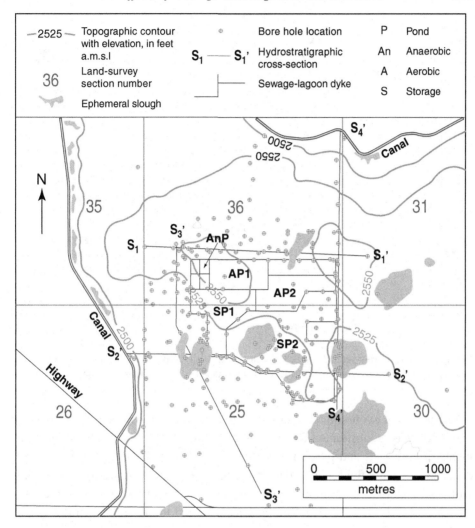

Fig. 5.18 Site map and preconstruction topography (Tóth, 2003, Fig. 8).

they can be considered as four distinct hydrostratigraphic units at the scale of the case. These are, in descending order (Fig. 5.19):

(1) a unit comprising glacial till, fractured coal, and bedrock shale, the 'shallow aquifer, A_1';
(2) fractured and discontinuous shale bedrock, the 'first aquitard, T_1';
(3) an extensive coal seam, the 'coal aquifer, A_2';
(4) unfractured shale bedrock, the 'basal aquitard, T_2'.

The top three units all bear signs of more or less strong mechanical disturbance by scraping and pushing of glacial ice. Fracturing, jointing, vertical displacements and thrusting of noticeably decreasing intensity with depth were observed in test pits and cut-off wall trenches (Fig. 5.20). Notwithstanding these indications, the site conditions were assessed by the consultant as suitable for lagoon construction even without lining the ponds (Fig. 5.21).

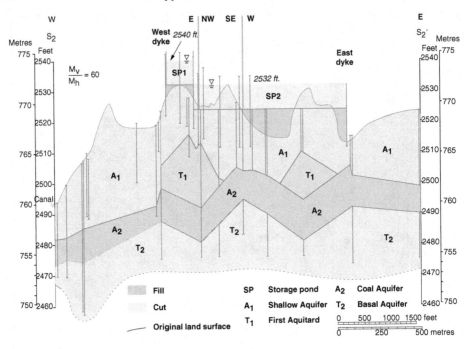

Fig. 5.19 'Cross-section S_2–S'_2' showing: hilltop position of lagoon, local hydrostratigraphy, and discontinuity of aquitard T_1 (Tóth, 2003, Fig. 8).

The temporal evolution of water-level changes (actually sewage-level) beneath the ponds was reconstructed for the topmost unit, aquifer A_1, based on measurements in boreholes and piezometers taken after filling of the lagoon begun. The data showed that a fluid mound developed under the lagoon upon commissioning, with spur-like lobes of contours pointing west–southwest thus indicating leakage beneath the dykes (Fig. 5.22). The flow has resulted in precipitated solids left by the evaporating effluents in the surrounding area (Fig. 5.23). Figure 5.24 presents the summary of the main causes of the lagoon's failure.

5.2.4 *Deep penetration of contaminants in recharge areas and saltwater up-coning in discharge areas, central Netherlands*

'In response to the pressing problems of appropriate water management, and the growing public and political concern about nature conservation and the need to safeguard ecosystems for future generations, the Dutch government formulated (the) project "National Analysis of Regional Groundwater Flow Systems" ' (Engelen and Kloosterman, 1996, p. 55). This project was launched in 1991 and conducted by 'TNO Institute of Applied Geoscience' (IAG). A standard component of the project's published results are cross-sections relating groundwater chemical composition to flow patterns.

Fig. 5.20 Fractured shale of aquitard T_1 (Tóth, 2003, Fig. 8, photo by Tóth).

- Surficial material: glacial till, $K \approx 10^{-9}$–10^{-7} ms^{-1}, "impermeable in engineering terms".
- The underlying shale bedrock: "impermeable".
- The only possible seepage would:
  "... follow the glacial till/bedrock interface..."
- Any seepage that might occur would reach the canal in ≈ 900 years.
- In summary
  "The hard till and shallow bedrock... excellent foundation for the construction of the sewage lagoon dykes...lining will not be required."

Fig. 5.21 Consultant's site assessment.

As part of the pilot phase of the project a section across central Netherlands was produced (Fig. 5.25a). Two relevant observations can be made in the present context: (i) 'Slightly polluted recently infiltrated waters' (Engelen and Kloosterman, 1996, p. 58) and 'An anthropogenic polluted groundwater and water infiltrated from the larger rivers' penetrate in recharge areas to depths reaching 100 m or more, while this depth is only 0–20 m in discharge areas (Fig. 5.25a); (ii) the interface between

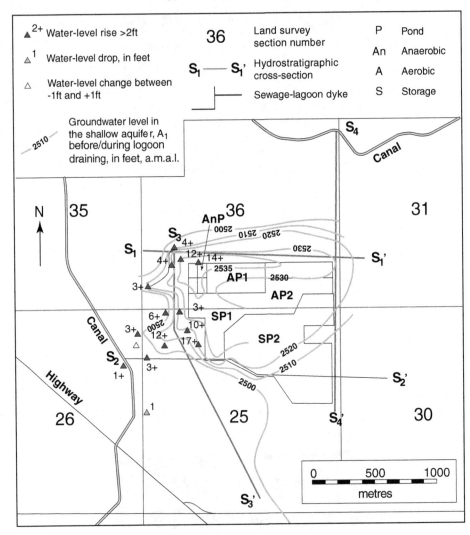

Fig. 5.22 Changes in water levels in Aquifer A_1 due to filling the lagoon and the resulting water-table mound (Tóth, 2003, Fig. 9).

the fresh- and the ubiquitous basal saline-groundwater is depressed to 350 m or more beneath the two principal groundwater recharge areas (Veluwe and Utrecht ice-pushed ridge complexes) but it is drawn to within 100 m to the surface by ascending waters in the major discharge area of Gelder Valley (Fig. 5.25a).

A noteworthy observation from the viewpoint of flow system analysis 'is the fact that the systems, recognizable in this section by the contrasting water qualities, are not much influenced by the presence of resistive clay layers. The development of the flow system underneath the river Vecht and Amsterdam-Rhine Canal (ARK) is

Fig. 5.23 Salt crust at site of seepage from the lagoon (photo by J. Tóth).

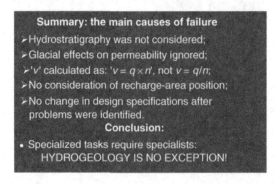

Fig. 5.24 Summary of lagoon problems and conclusion by the author.

apparently not impeded by the presence of a thick clay layer at a depth of 50–75 m'
(Engelen and Kloosterman, 1996, p. 59).

A schematic version of the E–W hydraulic cross-section of the central Netherlands (Fig. 5.25a) is presented in Figure 5.25(b) illustrating the Dutch terminology
and conceptual framework of flow-system classification (Engelen and Kloosterman,
1996, Fig. 8.3., p. 61).

5.2.5 *Transport of phosphorous by groundwater into Narrow Lake from near-shore recharge areas Alberta, Canada*

As one project in a series of studies aimed at evaluating the role of groundwater
in eutrophication of lakes in northern Alberta, Shaw *et al.* (1990) investigated the
possible sources and transport routes of phosphorous into Narrow Lake. 'Data

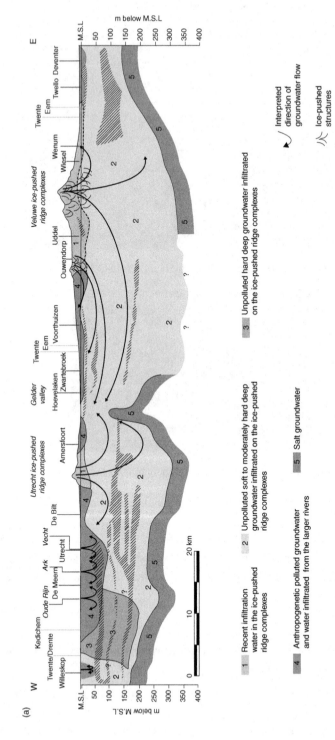

Fig. 5.25 West–east cross-section through central Netherlands, showing the major genetic types of groundwater: (a) flow pattern inferred from water levels and chemical water types; (b) schematized version of flow pattern to highlight the structure of the regional flow systems and system boundaries (modified after Engellen and Kloosterman, 1996, Plate 9 and Fig. 8.3, p. 61, respectively).

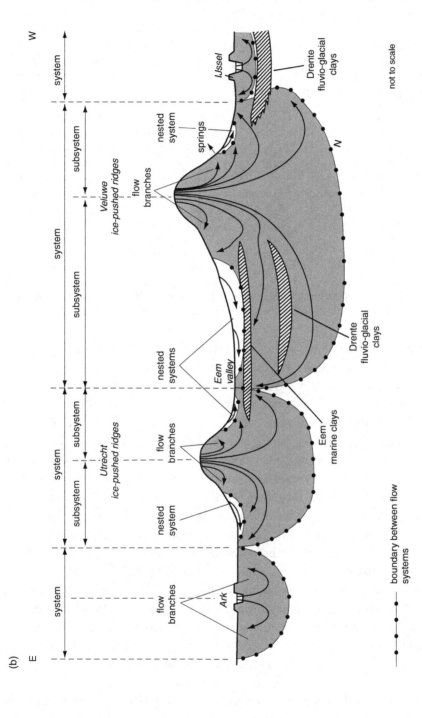

(b)

Fig. 5.25 (cont.)

not to scale

boundary between flow systems

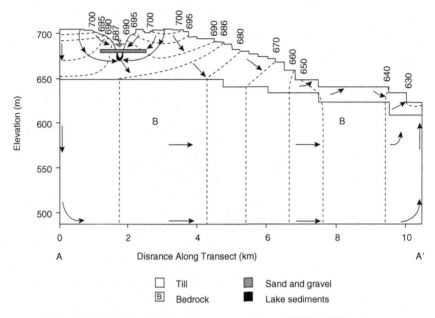

Fig. 5.26 Modelled groundwater flow-pattern along NNW–SSE transect across Narrow Lake, north-central Alberta, Canada (Shaw *et al.*, 1990, Fig. 5, p. 880).

from a drilling programme, major ion concentrations, environmental isotopes, and computer simulations indicated that the lake gains water through the nearshore region from a small, shallow groundwater flow system; at deep offshore regions, water moves from the lake to the groundwater flow system' (Shaw *et al.*, 1990, p. 870). (The above conclusions were corroborated by direct measurements of groundwater seepage on the shore and the lake's bottom.) Figure 5.26 shows a groundwater flow pattern modelled along a transect across the lake. The lake itself is situated in a regional recharge position but its deeply incised narrow valley induces intensive shallow local flow systems to its banks. Although contributing 'less than a third of the total water load to the lake, groundwater may be the major single source of P to epilimnic lake water. Groundwater influx occurs near the shore, so that P transported by groundwater enters the epilimnion, where it could be directly utilized by lake biota' (Shaw *et al.*, 1990, p. 885).

Although the above cited paper makes no mention of it, the authors have suggested, through personal communication, that cesspools, outhouses, and garden fertilizers associated with recreational cottages around the lake may be the source of phosphorous, as well as of nitrogen, providing yet another example for increased vulnerability of recharge areas to groundwater contamination.

5.2.6 *Cause and reclamation of liquefied ground, Trochu, Alberta, Canada*

After a few years of above normal precipitation in the early 1970s, the outbreak of numerous 'soap holes' and soft and wet soil patches made cultivation of approximately 65 ha of agricultural land difficult in the vicinity of the town of Trochu, Alberta. That piece of land, including the adjacent country road, located in 'SE 1/4, Section 28, Township 34, Range 24' (location identified according to the Canadian national land-survey system) and its surroundings had been inferred earlier from groundwater field manifestations as an area of local groundwater discharge (Tóth, 1966a, Fig. 24, p. 46). With the consent of the land owner and material contributions from the municipality responsible for road maintenance, a test-drilling and extended pump-testing experiment was conducted to determine if the liquefied- and saline-soil conditions could be ameliorated. The experiment was based on the flow system conceptualized for the area on the basis of observed groundwater-related field phenomena.

The flow system was thought to originate on a local topographic high at an elevation of approximately 930 m, with the discharge area 2.4 km to the south at 892 m, i.e. 38.5 m lower (Fig. 5.27).

The recharge-type characterization of the highland's general area was based on: (i) observed large seasonal variations of the water table in a local depression on the hill (i.e. a phreatic belt exceeding 6 m, when the depth of surface water in the slough is converted to subsurface water levels assuming 20 % porosity; Fig. 5.28); (ii) salinity-free soils; (iii) the presence of xerophytes beyond the borders of the local drainage centre.

The test drilling found an extensive horizontal and highly-permeable sandstone stratum at a depth of 62 m below the slough and at 23 m below the soap holes. This aquifer facilitated increased flow from the recharge- to the discharge-area but was confined by overlying more than 20 m thick clay layer. Water levels rose up to 3 m

Fig. 5.27 View south over local drainage basin from its recharge area to low-lying and distant (≈2.4 km) discharge area, Trochu, Alberta (photo by J. Tóth).

(a)

(b)

Fig. 5.28 Seasonally wet- or water-filled topographic depression on top of recharge area (located ≈200 m in opposite direction to Fig. 5.27) showing large seasonal changes in water levels: (a) deep (≈0.75 m) water body in 'slough' after snowmelt in May; (b) dry slough bottom and deep water table (>2 m below surface) in October (photo by J. Tóth).

above the land surface in observation wells completed in the aquifer near the soap holes (they were monitored during pumping in 4 m high open stand-pipes).

The pumped well, itself a flowing one, was completed in the confined aquifer and cased through the clay above it. The objective of pumping was not to dewater per se, but rather to reduce the pore pressure in the aquifer to the point that the vertical hydraulic gradient is reversed near the land surface from upward- to downward-oriented drive (i.e. from pointing downward to pointing upward; see comments regarding 'hydraulic gradient', Eq. 1.3a,b). This was accomplished by keeping the operating water-level constant at an approximate depth of 3 m below land surface. The soap holes started to collapse within two weeks after pumping started (Fig. 5.29). Thus the experiment succeeded in proving that liquefaction is the consequence of superhydrostatic hydraulic gradients driving water upward. If they are reduced to values that are equal to, or lower than, hydrostatic, the associated quick conditions end too.

Simultaneously with the hydraulic experiments, soil samples were taken periodically to see if an expected reduction of salinity could be demonstrated. For reasons beyond the experimenter's control, the soil-salinity study could not be completed. With the convincing, and expected, soil mechanical effect of the induced gradient reversal, however, there is reason to believe that an inversion of the hydraulic gradient would also have resulted in downward leaching of the salts, thus causing a reduction of soil salinization of the site, i.e. in chemical reclamation of the soil.

5.2.7 Groundwater flow and heat-flow anomalies: assessing low-enthalpy geothermal potential, northern Switzerland

Subsequent to the completion of the heat-flow density (HFD) mapping programme of Switzerland (Bodmer, 1982), a study of possible relations between heat-flow anomalies and regional groundwater flow was conducted in northern Switzerland by Bodmer and Rybach (1984, 1985). The study was part of a broader multidisciplinary investigation with the intent 'to assess the low enthalpy geothermal energy potential of the high HFD areas' (Bodmer and Rybach, 1985, p. 233). The general justification presented by the authors for their study approach was: 'In order to interpret thermal conditions at deeper levels of the Earth's crust by HFD maps it is essential to understand the effects of subsurface water circulation systems on the HFD field' (Bodmer and Rybach, 1985, p. 233). As their working hypothesis, they attributed the high heat-flow anomalies to heat transported by ascending gravity-driven groundwater. The validity of this hypothesis was to be assessed by the degree of correlation of positive and negative HFD anomalies with upward, downward directed groundwater flow, respectively.

Two areas with prominent HFD-anomalies were singled out for detailed investigations. These were the areas of 'Molasse of St. Gall', of approximately 2200 km^2

Fig. 5.29 Soap hole (quick soil) in discharge area: (a) mound and active discharge of water and mud before experimental pumping from confined aquifer began at depth of approximately 23 m; (b) same soap hole imploded in response of groundwater level drawn down to and held constant at ≈3 m below land surface (photos by J. Tóth).

in areal extent, and the 'East end of the Jura', approximately 2625 km^2, areas '1' and '2' respectively, in Figure 5.30.

In the 'Molasse of St. Gall' test area the recharge and discharge regions were inferred from parameters indirectly related to regional groundwater flow, namely: water chemical composition; deuterium D content (to infer topographic altitude of infiltration areas); tritium T content (to infer age); and geothermometers: silica, and Na/K/Ca of spring- and well-water samples (to infer reservoir temperature, i.e. depth of water penetration). High HFD values in the 'Molasse of St. Gall' area were found commonly to correlate with: (i) topographic lows (at or near valley bottoms); (ii)

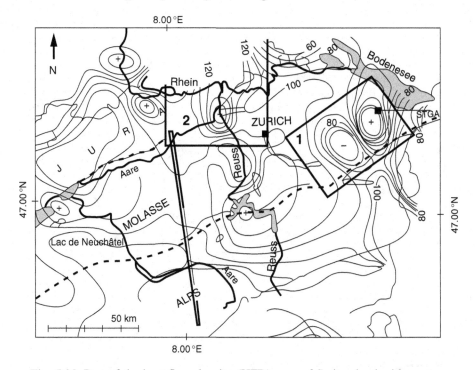

Fig. 5.30 Part of the heat flow density (HFD) map of Switzerland with contour interval of 10 mW/m^2. Rectangles '1' and '2' delimit, respectively, the 'Molasse of St. Gall' and the 'East end of Jura' HFD anomalies. The N–S double line is the trace of the cross-section in Figure 5.32. Dashed lines represent the boundaries of the geologic provinces Alps, Molasse and Jura (modified from Bodmer and Rybach, 1985, Fig. 1).

elevated reservoir temperature values (>25 °C) inferred from geothermometers; (iii) high infiltration altitude as indicated by δD values; (iv) low or zero tritium contents.

The 'East end of the Jura' test area contains a major positive HFD anomaly with a maximum value of 150 mW/m^2 and several thermal springs (Fig. 5.31). An 85 km long 2D heat-transport model between the Alps and the Aare River shows the calculated thermal effects of regional groundwater flow (Bodmer, 1982; Figs. 5.30 and 5.32), namely, the isotherms are depressed by recharging cold waters in the Alps and pushed upwards by warmed up discharging waters in the area of the Aare river.

Because the modelled regional system did not explain the local HFD anomalies, a local 2D heat-transport model was also constructed to simulate the subsurface temperatures around the thermal spring of Schinznach (Fig. 5.31). The W–E oriented section started 4 km west of the Aare River, in the assumed zone of local recharge (Fig. 5.33). The modelled HFD values at the upper boundary of the system are shown as the solid line and the field data as open triangles in the figure. The fit of the model to the field values was deemed 'reasonable' in the eastern part of the profile. The data scatter west of the discharge's focus (the thermal spring) is

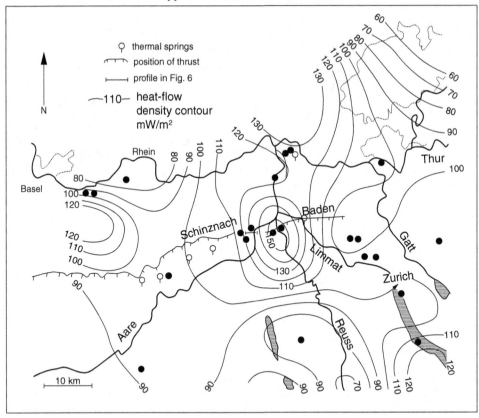

Fig. 5.31 Pronounced high heat flow density anomaly in the 'east end of the Jura' test area associated with several thermal springs. Stippled line indicates zone of enhanced permeability due to tectonic deformation (modified from Bodmer and Rybach, 1985, Fig. 4).

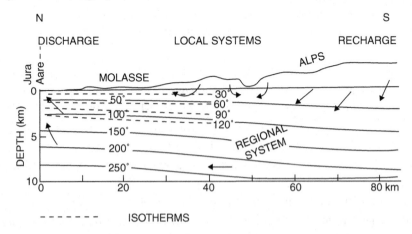

Fig. 5.32 Simulated isotherms (solid lines) for a deep water-circulation system along the N–S section trace in Figure 5.30. Dashed lines represent field measurements of subsurface temperatures (modified from Bodmer and Rybach, 1985, Fig. 5).

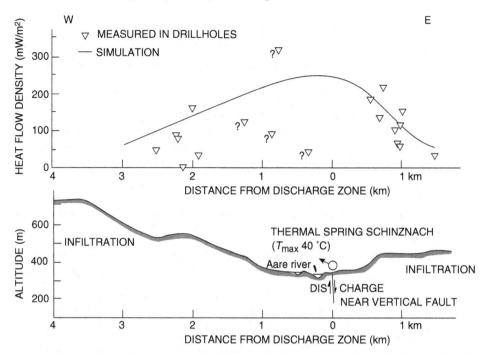

Fig. 5.33 HFD profile of a local E–W circulation system near the thermal spring of Schinznach. The triangles represent measured HFD values; the solid line is the modelled HFD at the upper surface of the sysem (Bodmer and Rybach 1985, Fig. 6).

attributed to differences between the modelled and actual topography (i.e. the relief of the water table) and to local disturbances such as infiltration of surface water, e.g., from the Aare River, within the groundwater discharge zone. 'Nevertheless, field observations, chemical geothermometers and mixing models (see NEFFF, 1980 in Bodmer and Rybach, 1985) substantiate this model and clearly indicate that in the case of northern Switzerland, small scale water circulation systems can be superimposed on the regional S–N drainage system' (Bodmer and Rybach, 1985, p. 242).

The conclusions of the study are best summarized in the words of the authors:

'– in regions with pronounced topographic relief and substantial water circulation, HFD mapping is commonly subjected to systematic errors if the HFD sites (drillholes) are located along valleys with relatively high HFD whereas the corresponding low HFD zones and the infiltration zones at the uplands remain undetected; these effects are not eliminated by the application of standard terrain correction methods;

– it is possible to obtain more information about the location of the low HFD anomalies and the size of the corresponding circulation systems by the application of geochemical methods including isotope analysis;

- in many cases several convective systems of different scale can interfere, making interpretation difficult. Thermally the most efficient convection systems are ones which operate on the km to 10 km scale;
- in constructing HFD isoline maps, the scale of the circulation systems and the size and location of the anomalies have to be taken into account;
- HFD sites which reflect local anomalies beyond the spatial resolution of HFD mapping need to be recognized and evaluated, in order to avoid errors by overestimating their effect on the regional HFD field' (Bodmer and Rybach, 1985, p. 244).

5.2.8 Increased susceptibility of slopes to failure in discharge areas: theoretical analysis

Based on rigorous mathematical analysis and numerical modelling, Iverson and Reid (1992) and Reid and Iverson (1992) have demonstrated quantitatively how gravity-driven groundwater flow influences the spatial distribution of effective stresses in slopes and how, in turn, these stresses affect the magnitude and distribution of the potential (i.e. susceptibility, tendency) for slope failure in hills and mountains. They have also shown the relations between the failure potential, on the one hand, and the fields of the water's seepage forces and the effective stresses generated by the combination of water flow and gravity acting on the poroelastic framework, on the other. The study's approach was novel in that it treated the effect of topography, i.e. of gravity, on groundwater flow simultaneously with the effect of the effective stresses. Previous studies analysed the influences of topography and subsurface stresses on groundwater flow separately (e.g., Tóth, 1963; Freeze and Whitherspoon, 1966, 1967; Savage *et al.*, 1985). The mathematical formulation and analysis were restricted to two-dimensional vertical sections of an infinitely extensive landscape with periodic topography (Fig. 5.34; Iverson and Reid, 1992, Fig. 1).

The susceptibility of a slope, or parts thereof, to failure is expressed by the spatial distribution of the 'Coulomb failure potential' Φ. It is a dimensionless measure of

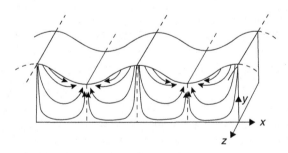

Fig. 5.34 Schematic block diagram of an infinitely extensive landscape with periodic topography. Streamlines show paths of gravity-driven groundwater flow (Iverson and Reid, 1992, Fig. 1).

the possibility of shear failure defined as $\Phi = |\tau'_{max}| / - \sigma'_m$, where Φ is the Coulomb failure potential, τ'_{max} is the maximum shear stress, and $-\sigma'_m$ is the mean normal stress, considered positive in tension (Iverson and Reid, 1992, p. 931). Φ is independent of the material strength and has a theoretical minimum of zero and a maximum of one. The effect of the flow of groundwater on the stability of a given slope can thus be assessed by comparing the failure potential fields calculated for the slope in dry and in saturated conditions. Figure 5.35 shows calculations for a straight hillslope (dry and saturated) composed of homogeneous material of specified parameters (Poisson's ratio $\upsilon = 0.333$; dry density $\rho_t = 1590$ kg/m^3; saturated density $\rho_b = 1990$ kg/m^3; water density $\rho_w = 1000$ m^3), and inclined at 26.6 degrees (2:1 slope).

The Coulomb failure potential in the dry hillslope is due to a uniform gravitational body force acting upon the (unsaturated) rock frame and was determined from the dry hillslope stress field (Fig. 5.35a; Iverson and Reid, 1992, Fig. 5). In this field (Iverson and Reid, 1992, Fig. 4, not shown here) the maximum near-surface compressive stresses roughly parallel the land surface while they align with gravity at depth. These stress conditions are reflected in the failure potential field Φ by the relatively uniform regions of high values occurring in a thin band subparallel and adjacent to the surface. Saturation of the hillslope with water induces a gravity-driven flow field $-\nabla h$ (where h is the normalized hydraulic head) which results in a seepage force field whose nondimensional distribution is depicted in Figure 5.35b; Iverson and Reid, 1992, Fig. 6). These seepage-induced body forces are maximum near the surface and negligible at depth, and they all have a down-slope oriented horizontal component. The vertical components of the seepage forces point downward at the hilltop and its vicinity and upward near the toe (i.e. opposite to the hydraulic gradient, see comment and explanation, Eq. 1.3a,b).

The combination of seepage forces with the gravitational and buoyancy forces (not shown here) results in the Coulomb failure potential field Φ displayed in Figure 5.35(c) (Iverson and Reid, 1992, Fig. 9). A comparison with Figure 5.35(a) shows that the values of failure potential in the saturated hillslope generally exceed those in the dry slope. In the saturated case the region of high Φ values ($\Phi > 0.5$) extends deeper and is shifted downslope, with maximum values near the toe where seepage is directed outward and Φ is approaching the theoretical maximum of 1. The net effect of the seepage forces induced by the gravity-driven groundwater flow on the Coulomb failure potential is seen in the difference in Φ between the dry and the saturated cases (Fig. 5.35d; Iverson and Reid, 1992, Fig. 10). Because of the horizontal flow components in the saturated case the greatest changes in Φ occur near the lateral boundaries of the flow domain. There, displacement of adjacent material cannot help distribute the stresses caused by groundwater flow due to the no-flow boundary emplaced to account for the periodicity of the topography. Significantly, the percentage increase in the failure potential is maximum near the toe. At the same time, that is also the area of the most intensive discharge of groundwater.

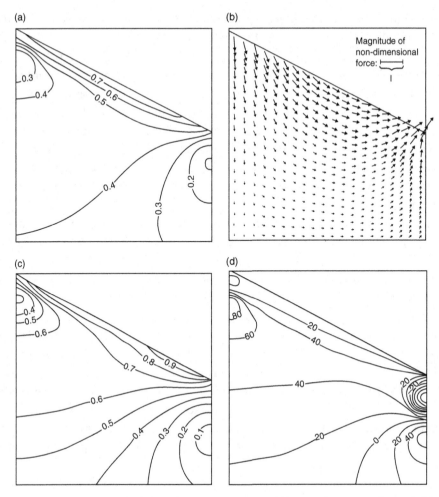

Fig. 5.35 Distribution of Coulomb failure potential Φ and seepage force field $-\nabla h$ in a homogeneous straight hillslope: (a) failure potential Φ in dry hillslope: (b) non-dimensional seepage force field $-\nabla h$ in saturated hillslope; (c) failure potential Φ in saturated hillslope; (d) percentage increase in failure potential Φ between the dry and saturated cases (modified from Iverson and Reid, 1992, Figs. 5, 6, 9, and 10, respectively).

A sensitivity analysis, in which the inclination of the slope was varied between horizontal-to-vertical ratios ($H:V$) of 3:1 (18.4°) and 1:1 (45°), found the patterns of groundwater flow, effective stress, and failure potential similar to the 2:1 $H:V$ ratio (26.6°) case, discussed above. In all these cases the Φ values are largest in the near surface regions, increase with the addition of gravity-driven groundwater flow, and reach their maxima near the toe of the slope, where flow is upward and outward.

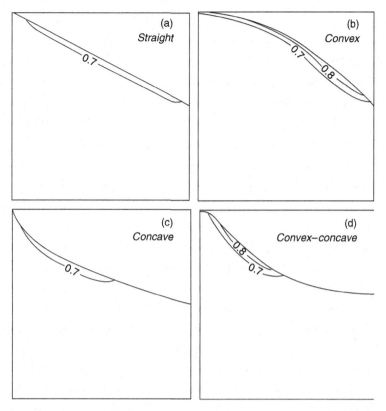

Fig. 5.36 Contours of failure potential Φ 0.7 and greater in dry hillslope with differing slope morphology (Reid and Iverson, 1992, Fig. 7).

In an attempt to approximate better the effect of natural landforms on slope stability, Reid and Iverson (1992) examined the patterns of failure-potential and seepage-force generated by four different slope profiles, namely: straight, convex, concave and convex–concave. The failure potentials Φ 0.7 and greater in dry hillslopes are shown in Figure 5.36. Both the distribution patterns and magnitudes of Φ change when the slope becomes saturated with flowing groundwater (Fig. 5.37). Again, in all cases, water flow results in lateral extension and deeper penetration of the areas of high failure potentials, as compared with the dry cases (cf. Fig. 5.36). The increases are linked to two principal factors: steepness of the slope and outward direction of the seepage-force vectors (discharge areas).

Heterogeneities in the hydraulic conductivity can significantly alter the groundwater flow pattern and thus affect the field of the Coulomb failure potential. Two simulated configurations of heterogeneity were a horizontal and a vertical layer in the otherwise homogeneous hillslope matrix analysed by Reid and Iverson (1992) representing, for instance, a sedimentary stratum and an igneous dike, respectively.

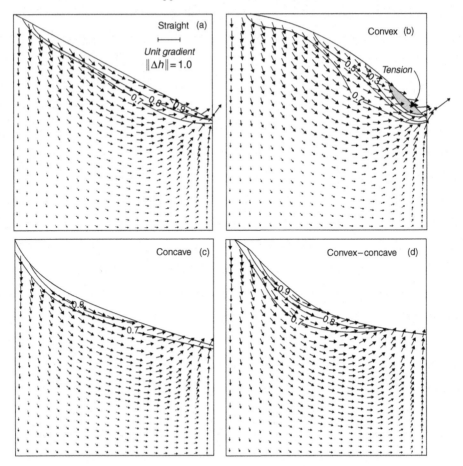

Fig. 5.37 Normalized seepage force vectors and contours of failure potential Φ greater than 0.7 in saturated homogeneous hillslopes with differing morphology (Reid and Iverson, 1992, Fig. 8).

The effects of both layers were evaluated with their permeabilities being higher and lower than that of the hillslope's mass. Results of the analyses are shown in Figure 5.38 (modified from Reid and Iverson, 1992, Figs. 13 and 14).

According to the numerical experiments, most of the changes in hydraulic head, related directly to the seepage forces by $-\rho_w g \nabla h$ and caused by contrasts in hydraulic conductivity K, occur between no contrast and contrasts of 3 or 4 orders of magnitude. Consequently, 'a hydraulic conductivity contrast of 4 orders of magnitude [$\log_{10}(K_{\text{layer}}/K_{\text{surrounding}}) = \pm 4$] is sufficient to obtain the maximum change in seepage forces that can be induced by hydraulic heterogeneities. These large contrasts can lead to regions of large negative pressure heads, indicating partially saturated conditions, in hillslopes with slope-parallel or horizontal layers' (Reid and

Iverson, 1992, p. 943). See also Fig. 3.12 and related explanation of 'subhydro-static' pressures in Tóth (1979, 1981); as well as Tóth (1962a), Fig. 8, concerning the asymptotically decreasing effect on potentiometric anomalies of increases in contrasts of permeability beyond 3 orders of magnitude between an ellipsoidal rock lens and its encasing matrix.

In the case of a horizontal layer with a K of one order of magnitude greater than the slope's matrix material (Fig. 5.38a) the flow-induced seepage forces are large near the crest but they are downward oriented thus not reducing the slope's stability. Indeed, due to the forces' orientation, slope stability increases there as indicated by the downslope shift of the 0.7 failure potential contour (i.e. the lower values move up) with respect to both the dry and the homogeneous saturated cases (Figs. 5.35a and 5.37a). The most vulnerable part of the slope is just below the layer's outcrop, i.e. in the discharge are a of the slope's matrix. The horizontal low-K layer divides the hillslope essentially into two hydraulically connected partial slopes: an upper small 'sub-basin' with a slightly permeable base, and a lower larger one with an impermeable base and receiving recharge from above through its low-K top (Fig. 5.38b). In this case, seepage forces area oriented outward from the total slope in two focalized areas: at the toe of the upper, partial slope, and at the toe of the whole slope. Owing to an intensification and, respectively, reduction of seepage forces at the higher and lower toes, the failure potential has also increased and reduced at these locations.

A vertical high-K layer located in the centre of the hillslope tends to intensify the discharge in the slope's lower half. Consequently, outward oriented seepage forces, and thus the potential for failure, are increased relative to both the homogeneous and the horizontal-layer cases (Figs. 5.37a, 5.38a, 5.38c). If, on the other hand, the hydraulic conductivity of the centrally located vertical layer is lower than that of the surrounding material (Fig. 5.38d) the hillslope is effectively divided into two laterally communicating 'sub-basins', both having their discharge areas at their respective toes. The seepage forces are outward oriented in both of these areas. Relative to the homogeneous case, these forces are enhanced in the higher discharge area but they are weakened at the lower toe owing to the damming, flow-reducing effects of the vertical barrier layer.

The principal conclusion of the Iverson and Reid (1992, p. 935) and Reid and Iverson (1992, p. 947) studies is that, in general, gravity-driven groundwater flow increases the Coulomb failure potential in the near-surface parts of hillslopes. The increases are particularly significant near the toe of a slope where seepage forces are strong and outward (i.e. surface-ward) directed. In these discharge areas the Coulomb failure potential Φ is higher than in corresponding parts of otherwise identical dry hillslopes. Recharge areas, on the other hand, have a similar Φ to that in the corresponding dry regions.

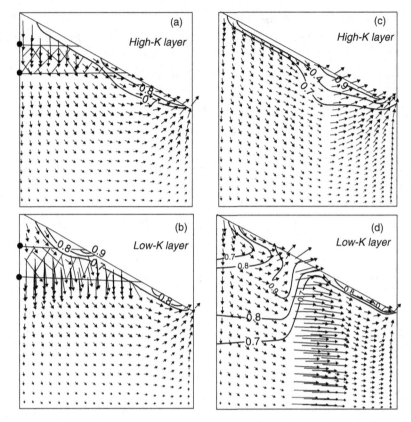

Fig. 5.38 Normalized seepage-force vectors and contours of failure potential Φ greater than 0.7 in a saturated hillslope. Horizontal layer: (a) with K value 1 order of magnitude higher than that of the slope's matrix; (b) with K value 1 order of magnitude lower than that of the slope's matrix (modified from Reid and Iverson, 1992, Fig. 13). Vertical layer: (c) with K value 4 orders of magnitude higher than that of the slope's matrix; (d) with K value 4 orders of magnitude lower than that of the slope's matrix (modified from Reid and Iverson, 1992, Fig. 14).

5.2.9 Analysis and mitigation of land-slide danger, Campo Vallemaggia, Switzerland

The general validity of the above conclusion is exemplified by the more than 200 year long history and many-faceted investigations of the Vallemaggia landslide in the southern Swiss Alps, crowned by the demonstrated success of the effect of a deep-drainage adit constructed to mitigate the propensity for continued mass movements (Bonzanigo *et al.*, 2007; Eberhardt *et al.*, 2007).

The body of the slide is up to 300 m deep and incorporates approximately 800 million m^3 of strongly disturbed and weathered metamorphic rocks that include schists, gneisses, metaperidotites and metacarbonates. It is divided longitudinally in the middle by a major subvertical fault into two distinct main bodies, the Campo and Cimalmotto blocks, transversally into several smaller blocks by subvertical fault zones of a few metres to tens of metres wide, and sub-horizontally along shear/slip surfaces at changes in lithology or in degree of alteration (Figs. 5.39a, b; Bonzanigo *et al.*, 2007, Fig. 5, respectively, Eberhardt *et al.*, 2007 Fig. 1).

Four main geomorphological zones can be distinguished on the slide's surface: (1) the head scarp, consisting of large intact blocks and open tension cracks where the slide separates from the stable intact bedrock; (2) the upper and central slide mass of highly disturbed large blocks with open transverse fractures between them, some being filled with weathered material and providing easy access to infiltration of precipitation into the subsurface; (3) the lower terrace of highly fractured weathered material comprising the area of greatest disturbance and deformation; (4) the slide toe, a four km long and locally 150 m high erosion front, composed of strongly fractured and weathered, poorly sorted soil-like material, and characterized by constant change because of the high rates of undercutting by the Rovana River at the base of the toe.

Two small villages, Cimalmotto and Campo Vallemaggia, are situated on top of the slide. Recorded observations of surface displacements go back for over 200 years and geodetically surveyed horizontal translation of approximately thirty metres was determined for the period between 1892 and 1995. Reoccurring movements of the slide have damaged roads, homes and historical buildings in Campo Vallemaggia, which was even abandoned temporarily in 1780.

The prime cause of the slope movements was attributed traditionally to the occasionally rapid and strikingly spectacular toe erosion and slope undercutting by the torrential Rovana River. However, based, among others, on his observations of several springs on the slide, the renowned Swiss geologist Albert Heim (1897) concluded that the sliding activity was due to the combination of the Rovana River attacking the slide toe, and the movements of the slope's upper sections induced by saturation of the slide mass (Heim, 1932).

The two conflicting views as to the principal cause of the sliding, namely, toe erosion or groundwater, could not be reconciled during the following more than 100 years. Nevertheless, measures were needed to slow or stop the movements in view of the precarious position of the two villages located on the lower parts of the slide mass. They kept moving closer to the edge of the 100 m high steep erosion front each year. As a result, a series of intensive investigations were conducted during 1983–1991, including geological, structural, and geomorphological mapping; seismic reflection and refraction surveys; borehole drilling; inclinometer and pore-pressure measurements; measurements of precipitation; hydrogeological and hydrochemical analyses; and surface geochemical monitoring.

(a)

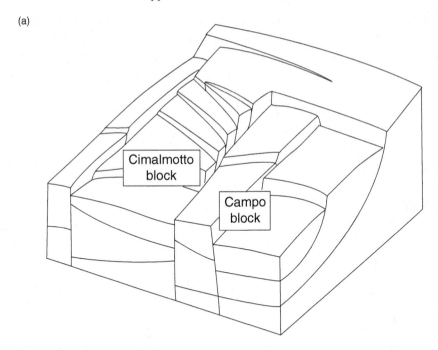

(b)

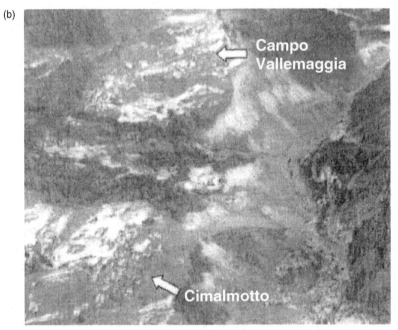

Fig. 5.39 (a) Three-dimensional block diagram showing subdivision of the Campo Vallemaggia landslide along fault zones and internal shears (Bonzanigo *et al.*, 2007, Fig. 5); (b) erosion scarp at the toe of the Campo Vallemaggia landslide and location of the villages of Cimalmotto and Campo Vallemaggia. Vegetated area in the middle left marks the approximate location of fault dividing the landslide into the Campo and Cimalmotto blocks (Eberhardt *et al.*, 2007, Fig. 1).

Of direct relevance to the objective of the present discussion are the systematically observed conditions of confined aquifers, 65 springs inventoried along fault lines, and hydraulic heads rising above land surface from deep boreholes in several areas of the slide's lower regions. Also, an existing spring blew in August 1991 on the NE bounding fault of the slide body, producing discharges of watery mud at rates above its normal flow. Based on analyses of the discharged water and mineral matter the phenomenon was ascribed to localized movement of the landslide mass compressing and forcing the mud from depth up to the surface along a fault. In another event, a period of intense precipitation in October 1993 provoked an acceleration of the slide mass and triggered two large landslides in the Campo block. This event as well as a more general correlation between precipitation, pore-pressure responses, and slide-block velocity, observed in one of the deep boreholes (CVM6), are shown in Figure 5.40. Note that the hydraulic-head value, 1390 m, is considered critical, below which only background sliding occurs.

Based on detailed analyses and interpretation of observed pore pressures, patterns of anisotropic hydraulic conductivity, fault-controlled flow paths, precipitation records, coupled hydro-mechanical behavior of the slide mass, and a 2 D groundwater flow model constrained by measured pore pressures, the authors concluded that the slide's 'behaviour is highly sensitive to accumulated pore pressures, resulting from long-term precipitation events. Results show that pore pressure values exceeding an apparent threshold value coincide with sudden acceleration of the slide mass. Velocities return then to background levels as the pore pressures dissipate and drop below threshold' (Bonzanigo *et al.*, 2007, p. 15).

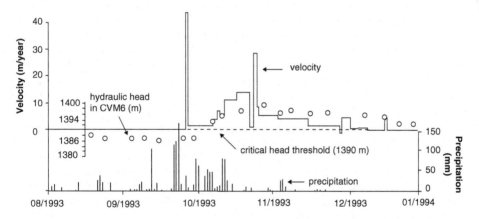

Fig. 5.40 Relationship between precipitation, pore-pressure response at depth (borehole no. CVM6), and slide-block velocity. Note: pore pressures are plotted in terms of hydraulic heads and superimposed on the velocity curve (Bonzanigo *et al.*, 2007, Fig. 22).

Notwithstanding the thoroughness of the hydrogeological investigations and the demonstrable conclusions derived from them, the competing view of the toe erosion being the mass movement's principal cause still had strong support. As a result, two different sets of proposed measures of stabilization were implemented, namely: (i) a 7 m diameter diversion tunnel to redirect the Rovana River away from the erosion front (constructed during 1993–1996); (ii) A deep-drainage adit driven below the unstable mass, with bore holes drilled upwards into the base of the slide's body (constructed during 1993–1995; Fig. 5.41).

During the adit's construction several hypotheses regarding the subsurface geology and hydrogeology were confirmed, such as: (i) numerous subvertical faults are present in which strong vertical flow gives rise to artesian conditions at the base of the sliding mass; (ii) the flow in these subvertical faults is obstructed horizontally by the central gouge's low permeability but enhanced vertically by the highly damaged zone of the adjacent rock; (iii) hydrogeochemical signatures of water collected from springs at the surface and within the drainage adit are comparable.

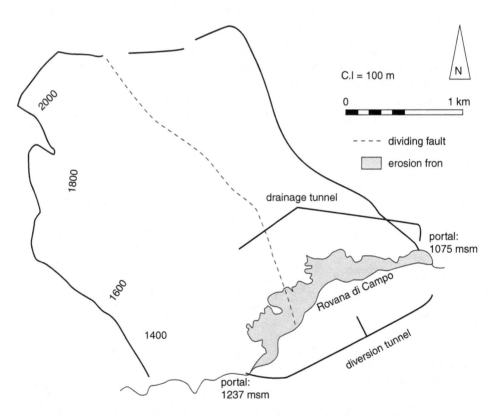

Fig. 5.41 Location of the Rovana River diversion tunnel and Campo Vallemaggia landslide drainage adit (Eberhardt *et al.*, 2007, Fig. 4).

As a consequence of the induced deep drainage, the naturally upward flow of groundwater through the unstable zone was redirected towards the adit instead of discharging in the lower portions of the hillslope and the valley bottom beneath the foot of the landslide as shown in Figure 5.42. The initial, approximately 1 km length of the adit was relatively dry. Within approximately 100 m ahead from the point of the southwesterly bend in the tunnel (Fig. 5.41) a water bearing fault was encountered with a hydraulic head of 1420 m above sea level. From this point on the discharge kept increasing reaching 9 l/s at the adit's completion. With the addition of 35 drainage bore holes the total discharge at the adit's portal grew to 50 l/s.

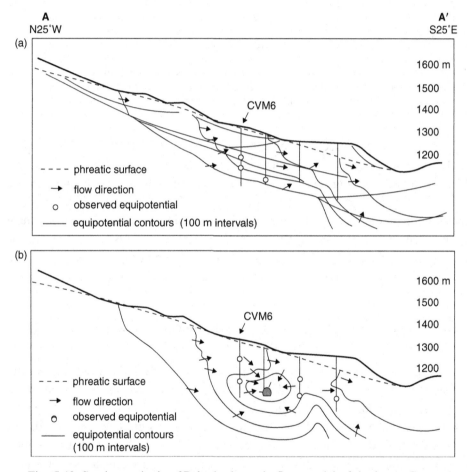

Fig. 5.42 Semi-quantitative 2D hydrodynamic flow model of the lower Campo block: (a) before and (b) after opening of the drainage adit, showing piezometric sites and observations with bore hole CVM6 marked, equipotential contours and groundwater flow vectors (modified from Eberhardt *et al.*, 2007, Fig. 9).

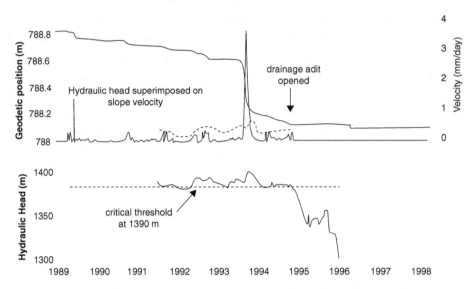

Fig. 5.43 Correlation between downslope velocities of the Campo block and pore pressures measured in borehole CVM6 before and after deep drainage. Pore pressures are expressed in hydraulic head (Eberhardt *et al.*, 2007, Fig. 11).

Not only was the effect on pore pressures of the drainage boreholes 'surprisingly immediate', but, in the authors' words: 'The response of the landslide was likewise immediate (Fig. 11). The reversal of the pore pressure gradient within the landslide transition zone was seen to quickly stabilize the Campo block through the subsequent increase in effective stress and resisting forces along its basal shear zone. Again, it was quite interesting to note the quickness of this process despite the overall low permeability of the crystalline rock and immense volume of material involved in the instability' (Fig. 5.43; Eberhardt *et al.*, 2007, Fig. 11).

5.2.10 *Groundwater flow systems and eco-hydrological conditions: study of the effects of land-use changes*

Eco-hydrology is a new interdisciplinary science that deals with the relationship between the type and quality of plant associations, on the one hand, and the chemical conditions and water availability at their site, on the other. The new science's emergence can be traced to the recognition by plant ecologists that 'gravity-driven groundwater flow influences the site conditions through the spatial distribution of nutrients and other relevant chemical agents. Especially upward seepage may produce and maintain site conditions that are essential for various relatively rare plant species and communities' (Klijn and Witte, 1999, p. 65). This recognition has inspired numerous studies based on, and cross-fertilizing, knowledge and ideas of ecologists, hydrologists and hydrogeologists, such as Wassen *et al.* (1990), Winter *et al.* (1998), Klijn and Witte, (1999), Batelaan, (2006), Zinko *et al.* (2006), Jansson *et al.* (2007). At the same time, it has also elicited the prospect of understanding, and even controlling, the effects of some land-use practices in areas that are linked to ecologically valuable and/or vulnerable tracts by groundwater flow systems. To illustrate the type and nature of possible eco-hydrological problems and approaches a study by Batelaan *et al.* (2003) was selected because of its strongly interdisciplinary nature, advanced methodology and employed techniques, quantitative characterization of gravity-driven groundwater flow systems and, importantly, its application to a major real-life practical problem.

The study area of Batelaan *et al.* (2003) is located in the drainage basin of the Grote-Nete river in northeastern Belgium. It covers 239 km^2 and its eastern half is part of the topographically relatively high Campine region (Fig. 5.44). Thirty per cent of the area is used for agriculture, the rest is covered by forests, grass land, some heather and open water, and about 18 per cent of it is built up.

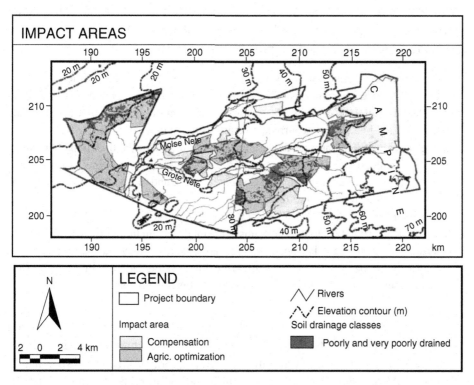

Fig. 5.44 Grote-Nete study area showing topography and impact areas designated for agricultural improvement and compensation measures (modified from Batelaan *et al.* 2003, Figs. 1 and 3, pp. 93 and 95).

In the framework of the 'Grote-Nete land-use project', agricultural productivity is intended to be improved by artificial drainage in certain tracts of land called 'optimization' areas. However, to counteract the expected reduction in groundwater recharge after drainage, and thus to prevent ecological damage in areas relying on groundwater discharge, measures are planned to reduce surface runoff and evapotranspiration in 'compensation' areas by means of closing ditches, installing weirs, and changing vegetation types.

Attainment of such objectives requires quantitative knowledge of the sensitivity of groundwater conditions (such as discharge rates, areal extent, water-table depth, water quality) in the discharge areas of flow systems to land-use changes in the same systems' recharge areas. In turn, the ability to evaluate that sensitivity requires knowledge of the existing conditions and the functional relation to their controlling factors which include, for instance, relief of the water table, distribution and strength of sources and sinks, rates of infiltration and evapotranspiration, and rock-framework permeability. Once the existing situation and its spatial and temporal dependence on the controlling factors are determined, both antecedent conditions and future responses to imposed changes can be estimated.

Batelaan *et al.* (2003) proposed a methodology to produce the information required for the planning and implementation of the Grote-Nete land-use project. The methodology comprises three principal components.

(i) Development of a steady-state numerical groundwater flow model for the study area, calibrated to field observations of water levels and hydrologic parameters (precipitation, surface runoff, base flow of streams, spring flow-rates, etc.), and assessed and complemented by mapped occurrences of phreatophytes.
(ii) Characterization of the present-day groundwater flow systems by their length, location, extent of recharge and discharge areas, throughflow times from points of recharge to points of discharge, hierarchical type (local, regional), and discharge rates, and by correlating discharge regions of different ecologies with flow systems of different types by cluster analysis.
(iii) Modelling of groundwater flow systems with modified input parameters in order to determine the hydrological and ecological conditions: (a) as they may have been before human activities had significantly disturbed the flow systems and (b) as they may become after the contemplated Grote-Nete land-use project is implemented.

(i) Development of a steady-state numerical groundwater flow model. The main elements of the groundwater flow model were the USGS' three-dimensional finite difference MODFLOW code (Harbough and McDonald, 1996) applied specifically to the Grote-Nete area (Batelaan *et al.*, 1996) and modified by (a) the SEEPAGE package of Batelaan and De Smedt (2004) to enable determination of the position of recharge and discharge areas, and (b) the DRAIN package of McDonald

and Harbough (1988) to estimate discharge rates under the local conditions. In addition, the spatial variation in recharge due to differences in land use, soil type, slope, groundwater level, meteorological conditions, and so on, was taken into account by a quasi-physically based methodology termed WetSpass ('Water and Energy Transfer between Soil, Plants and Atmosphere, under quasi Steady State'; Batelaan and De Smedt, 2007). The WetSpass code estimates surface runoff and evapotranspiration from slope, soil type, land use, intensity of precipitation, potential evapotranspiration, soil-moisture storage capacity and soil cover. Calibration of the integrated code was based on 'comparison of observed and calculated water levels as well as on the surface and groundwater balance of the basin' (Batelaan *et al.*, 2003, p. 90).

(ii) Characterization of the flow-systems. Characterization of flow systems started with the delineation of the modelled discharge areas. This step was based on the contiguity of computed discharge regions and on the similarity of vegetation types in those regions. Locations where groundwater discharge should occur were recognized by the simulated water table reaching or rising above the land surface, and by calculated recharge values being negative. Most of the simulated discharge points appeared in bands of $\approx$500 m width along the main water courses. Discharge was identified over approximately 16 per cent of the project area with an average rate of 4 mm/day and with narrow strips of 5 mm/day fluxes limited to the thalwegs of the valleys. Calibration of the model to water levels in 38 piezometers in the area resulted in a correlation coefficient of $R^2 = 0.99$. Hydrologically, the model was verified by comparison of the calculated runoff to 10 years of discharge data from a gauging station downstream from the area. The observed specific discharge of 329 mm/year compared within 2 per cent with the simulated discharge.

An extensive programme of vegetation mapping was conducted for two principal purposes:

(a) to verify and refine the position and extent of the simulated discharge regions;

(b) to detect possibly different eco-hydrological characteristics in different discharge areas and to seek correlation, if any, between such characteristics and different types of flow systems.

Twenty-three phreatophyte species (plants obtaining their water needs from the saturated zone) were selected as potential indicators of the presence of discharge and the quality of groundwater. The expected indicator value of a phreatophyte, or a phreatophyte community, is based on its preference to specific conditions of chemical quality, nutrient content (trophic level), and abundance and temporal variability of its water supply, on the one hand, and the differences in these site attributes between the recharge areas and discharge areas of the given flow system (Chapter 4, Fig. 4.1). Out of the numerous aspects of water quality, alkalinity and trophic status of a site were chosen as significantly different in recharge and discharge conditions, and to which phreatophytes would be sufficiently sensitive in their indicator role. Average groundwater alkalinity was found to correspond to a pH of approximately 6 in the study area, with slightly lower values towards the recharge-, and higher values towards the discharge-areas. The trophic levels generally increase in the direction of flow from atmotrophic (nutrient poor, similar to atmospheric precipitation) to lithotrophic (enriched by flow through the rock framework). Through random visits in the area's valleys, 193 phreatophytic sites were identified ranging in extent from 5 to $100\,000\,\mathrm{m}^2$, averaging $4500\,\mathrm{m}^2$. From the mapped locations of phreatophytes, 79 per cent were found to lie within discharge areas as calculated by the model and most of them had medium or high discharge rates. The 'mismatches' were attributed to scale limitations and uncertainty in the phreatophyte mapping.

Eighteen different discharge regions were delineated based on the criterion of areal contiguity of discharge locations as predicted by MODFLOW's SEEPAGE-DRAIN package and refined by the results of phreatophyte mapping. The recharge areas associated with the individual discharge regions were determined by forward particle tracking using Pollock's (1994) MODPATH code. The eighteen flow systems are shown in Figure 5.45 (Batelaan *et al.*, 2003, Fig. 10) along with their contributing recharge areas and groundwater travel times from recharge to discharge locations within individual groundwater flow systems. Major additional characteristics of the flow systems, keyed to the map through the number of their discharge areas, are summarized in Table 5.2 (Batelaan *et al.*, 2003, Table 2).

One of the goals of the study was to evaluate possible relations between the ecological status of discharge areas, on the one hand, and the type of flow systems feeding them, on the other. Such an inquiry requires the ability to make comparisons among ecologies of individual discharge-areas as well as among flow systems or, in other words, to classify them using characteristic parameters. To this end, a cluster analysis (Seyhan *et al.*, 1985) was conducted based on comparison of plant species, alkalinity, and trophic level of the discharge areas, and average flow time from recharge- to discharge-areas, ratio of the extents of recharge areas to discharge areas, and discharge flux of the flow systems. According to the analysis, all the eighteen discharge areas could be classified into one of four distinct clusters, identified by the Roman numerals I–IV.

The discharge areas of Cluster I (regions 1, 3, 11, 13 and 14; Fig. 5.45, Table 5.2) and Cluster II (regions 2, 4, 17 and 18; Fig. 5.45, Table 5.2) are all located in headwaters of the study area's highest locations. Their flow times are similar and indicate local but deep groundwater systems by the fact that their recharge areas extend to the regional groundwater divide. The alkalinities are similar and lower

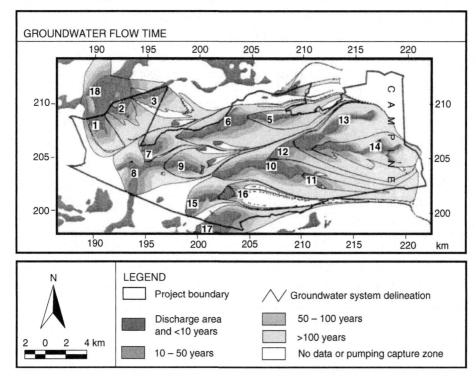

Fig. 5.45 Simulated groundwater flow systems for the present: numbers 1–18 indicate different discharge areas; their contributing recharge areas are delineated by thin black lines; and groundwater travel time from each recharge location to the indicated discharge location within a groundwater flow system is shown by grey-scale contours (Batelaan *et al.*, 2003: Fig. 10).

than in the other clusters, and indicate atmotrophic, i.e. low pH, water. Cluster I has a higher recharge- to discharge-area ratio, which reflects the different geomorphic positions of the two groups' discharge regions. In summary, Cluster I is characterized as local but deep flow systems situated upstream and carrying atmospheric water. Cluster II flow systems are local, shallow, situated downstream and carry atmospheric water also.

All average parameter values for Clusters III (regions 5, 6, 10 and 12; Fig. 5.45, Table 5.2) and IV (7, 8, 9, 15 and 16; Fig. 5.45, Table 5.2) are higher than for Clusters I and II (except for flow time and flux for Cluster IV) indicating their character as more regional and more lithotrophic (higher nutrient values). Cluster III comprises the most regional flow systems with long flow times, high recharge- to discharge-area ratios and high seepage fluxes. Cluster IV systems are similar to those of Cluster III except for their more downstream location and shallower penetration depth, which explains their shorter throughflow times.

Table 5.2. *Characteristics of each delineated groundwater flow system for the present situation: size of recharge and discharge areas, average groundwater discharge, and average groundwater flow time (Batelaan et al., 2003, Table 2)*

Region	Recharge area (km^2)	Discharge area (km^2)	Average discharge (mm/day)	Average flow time (years)
1	5.1	1.5	3	68
2	2.4	1.8	2	24
3	5.8	1.3	4	69
4	2.8	1.4	2	216
5	9.7	2.0	5	190
6	25.5	5.2	5	208
7	8.0	1.9	4	20
8	16.1	3.6	4	114
9	9.2	2.1	4	148
10	33.9	7.1	5	220
11	3.8	1.0	4	94
12	8.6	1.6	5	227
13	26.4	5.8	5	239
14	16.0	4.4	4	184
15	8.4	2.3	4	132
16	4.2	1.1	4	146
17	11.7	5.2	3	120
18	27.6	12.5	3	154

(iii) Modelling of groundwater flow systems with modified input parameters. In order to get a feel for the sensitivity of the discharge-dependent ecological conditions to changes imposed on the flow-systems, as well as to determine the expectable environmental impact consequent upon implementation of the Grote-Nete land-use planning project, the groundwater model was re-simulated with modified input parameters to evaluate two different scenarios, namely: (a) 'pre-development', i.e. representing conditions that existed prior to any significant human influence, and (b) 'future situation', i.e. showing the effects of possible future changes in land use due to the Grote-Nete project. The calculated effects are expressed as percentage changes relative to the characteristics of the presently existing, i.e. initially evaluated, eighteen groundwater flow-systems shown in Table 5.2. (The effects of these modifications are summarized in Batelaan *et al.*, 2003, Tables 3 and 4.)

In the pre-development scenario all groundwater abstractions were ignored and recharge was calculated without agricultural and urban land use. As a consequence of these imposed changes, both the discharge- and recharge-areas expanded by, respectively, 7.9 and 10.2 per cent, indicating a considerable increase in the combined surficial areas of the natural flow of groundwater. The increase is attributed to the absence of groundwater abstraction prior to development of groundwater resources. Among the eighteen flow systems shown in Figure 5.45, however, the changes vary widely, between 128.6 and −18.4 per cent for recharge areas, and from 181.8 to −11.4 per cent for discharge areas. The change in flux, as compared with the present situation, varies between 13.9 and 19.5 per cent, while the total discharge increases in fourteen of the eighteen regions ranging from 3.2 to 146.7 per cent. It is interesting to note that while region 16 has the largest increase in total discharge (146.7 per cent, attributed to the absence of abstraction by wells) its flux of discharge is reduced by a change of −10.8 per cent. The same phenomenon also occurs in regions 1, 5 and 12. These changes are explained by a percentage wise larger increase in the area of discharge than in the flux: the increased total discharge is thus spread over a larger discharge area than the present one.

Thus it appears that almost 80 per cent of all regions used to receive larger amounts of discharge in the pre-development state than today. In approximately 25 per cent of the cases the difference was due to an increase in the area of discharge while the flux decreases. The reduction in flux can adversely affect groundwater-dependent vegetation because lower fluxes may result in larger seasonal fluctuations of groundwater levels as well as in temporary desiccation of the site.

Simulation of the groundwater flow systems for the future situation was conducted with decreased recharge in the intended areas of agricultural 'optimization' and increased recharge in the planned 'compensation' areas. As a result, discharge has increased in one-third, decreased in one-third, and remained unchanged in one-third of all discharge regions combined. Some of the reductions in discharge were attributed to the artificial drainage in recharge areas causing reduced infiltration. Similar to the pre-development scenario, the intensity of local discharge does not necessarily increase just because the total discharge increases.

The changes in discharge flux and in the size of recharge- and discharge-areas are generally smaller than in the pre-development scenario. The difference has been attributed to the compensating effects of negative and positive changes in the planned land uses on the groundwater systems. Changes in groundwater levels vary from a maximum increase of 0.55 m to a maximum decrease of 0.07 m. In general, the effects of the totality of compensating measures seem to balance quantitatively those of the optimization measures. However, large local changes in hydrologic conditions may occur due to spatial differentiation of the measures. Consequently, the ecological effects of the proposed land-use changes can be expected to vary strongly on the local scale. Notwithstanding large changes in discharge fluxes and discharge-area extents, the decline of groundwater levels is relatively small. Hence the implied change in recharge as a consequence of land-use changes is regarded as mild.

In summary, the Batelaan *et al.* (2003) study presents a methodology for the quantitative characterization of gravity-driven groundwater flow systems by numerical modelling verified by field data and combined with mapping of phreatophytic plant communities growing in discharge areas. Evaluation by cluster analysis of the functional dependence of groundwater discharge, thus eco-hydrologic conditions, on flow-system types has allowed modelling of the effects of changes in land use on these conditions. Land-use changes were represented by changes in the model's input parameters. The methodology was applied to the Grote-Nete land-use plan project in northeastern Belgium. The plan's objective was to improve productivity in certain sectors of the project area by artificial drainage while preserving ecologically valuable wetlands. Preservation would be ensured by maintaining the natural groundwater supply to the protected areas by increasing recharge through compensatory land-use changes. It is concluded that modelling of groundwater flow-systems is a requisite for the evaluation of the intricate distribution of and changes in location, size and fluxes of the discharge areas.

5.3 Site-selection for repositories of high-level nuclear-fuel waste: examples for groundwater flow-system studies

The isolation of high-level nuclear-fuel waste, or high-level radioactive waste, has become perhaps the globally most studied and debated practical problem in the applied earth-sciences during the last five decades. One of its focal questions has concerned the role of gravity-driven regional groundwater flow. The final and firm conclusion is that the understanding, knowledge, and exploitation of the distribution of regional groundwater flow are indispensable in the search for and selection of suitable repository sites. In Proceedings of Workshop W3B, *Geological Problems in Radioactive Waste Isolation: a World Wide Review*, 10 out of 19 of the submitted national reports placed explicit emphasis on the importance and methods of characterization of gravity-driven groundwater flow (Witherspoon, 1989). Two main reasons appear to be the basis for the interest. First, gravity-driven groundwater flow has come to be recognized as the single most important, if not the only, ubiquitous mechanism of subsurface transport of matter and heat beneath the Earth's terrestrial areas. Second, gravity is the common water-driving force in the 500–1000 m depth range, which is considered by the majority of project planners as the most suitable, and thus most probable, realm for future repositories.

It is instructive to note, however, that while hydrogeology in general, and flow-system studies in particular, constitute a major part of current investigations in the search for and characterization of potential repository sites, the importance of groundwater's role in the problem was recognized slowly and accepted only reluctantly in the 1960s and 1970s. It is tempting also to speculate whether the progress in thinking and attitude was not due to the flow-system concept itself, which had its principal period of evolution and maturation exactly in those two decades. Initially, engineered barriers were considered as the chief, if not the only, means of radioactive waste isolation. The thinking was changed gradually by the increasing number of new-thinkers who joined the technical teams and advisory boards of the various waste-disposal projects. Annotated relevant aspects of high-level nuclear-fuel waste-isolation programmes are presented briefly below from Canada, Sweden, Switzerland and the USA.

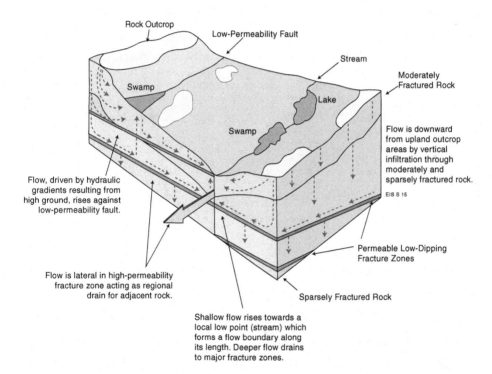

Fig. 5.46 Schematic portrayal of groundwater flow-system types in the Canadian Shield landscape: 'A disposal vault would be located so that the characteristics of the local and regional groundwater flow systems and the groundwater chemistry could be used to best advantage in the design of the disposal vault' (AECL, 1994b, p. 14).

5.3.1 Canada: the Recharge Area Concept (AECL: Atomic Energy of Canada, Ltd)

In a joint statement, the governments of Canada and the province of Ontario directed Atomic Energy of Canada Limited (AECL, a 'Crown Corporation', i.e. a state owned company) in 1978 to develop a concept of deep geological disposal of nuclear fuel wastes. Search and selection of an actual disposal site would not begin until after a full public hearing at the federal level and approval of the concept by both governments took place.

AECL submitted an 'Environmental Impact Statement' (EIS) in 1994, on the methodology, proposed implementation, and expected environmental consequences of a nuclear waste repository to be built in the Canadian Shield. The EIS comprised nine 'primary reference' documents, one condensed volume reviewing the primary references (AECL, 1994a), and a brief summary document (AECL, 1994b). The documentation contained the distilled but detailed results produced

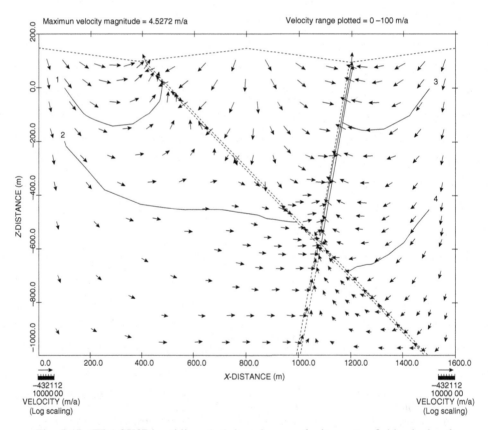

Fig. 5.47 'TRACK3D' pathlines 1–4, based on a velocity vector field calculated by AECL's MOTIF computer code (Nakka and Chan, 1994, Fig. 16, p. 100).

and presented in hundreds of Technical Reports and other types of scientific publications during the preceding 16 years of investigations at a cost of over Can$ 400 million ($\approx$US$ 250 million in the 1990s). One of the principal areas of research in the entire programme was hydrogeology, in general, and groundwater flow systems, in particular. Based on extensive field investigations in several research areas, the study showed that groundwater is driven by gravity and moves in hierarchically nested flow systems in the crystalline rocks of the Canadian Shield with their patterns modified by faults and fractures (Fig. 5.46).

The empirical observations were used by AECL as the theoretical basis for the development of a particle-tracking code (TRACK3D) to compute flow paths of conservative (i.e. non-reactive) contaminants. To calculate real-life path lines of transported contaminants TRACK3D used flow-velocity fields computed for site-specific situations by other codes of groundwater flow (Fig. 5.47).

In spite of its own field observations and model results, as well as the insistence of its own 'Technical Advisory Committee' (of which the present author was the hydrogeologist member), however, AECL was reluctant to acknowledge the inherent advantages of a recharge-area repository-location, possibly in order to keep its options open for the future in site selection. In view of this reluctance, the Technical Advisory Committee wrote in its 12th Annual Report (TAC, 1992, p.10):

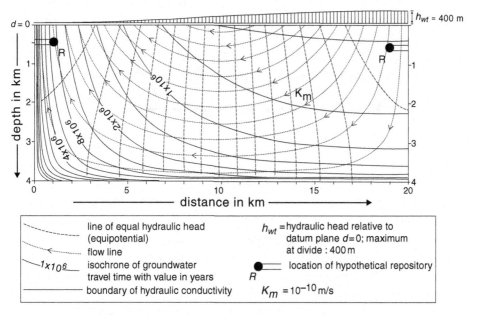

Fig. 5.48 Calculated distributions of hydraulic heads, groundwater flow lines, and isochrones of groundwater travel-time between water divide and valley bottom in a basin of homogeneous matrix hydraulic conductivity (Tóth and Sheng, 1996, Fig. 3, p. 11).

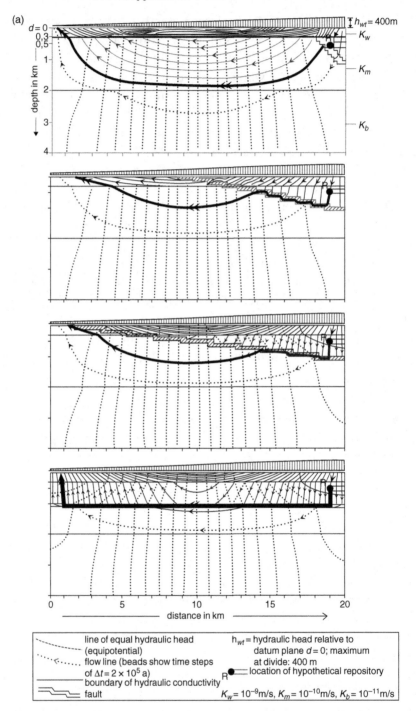

Fig. 5.49 Flow trajectories to and from a repository as a function of attitude of hypothetical fault in a basin of stratified rock framework: (a) repository in recharge area; (b) repository in discharge area (Tóth and Sheng, 1996, Figs. 6, p. 16 and 8, p. 18, respectively).

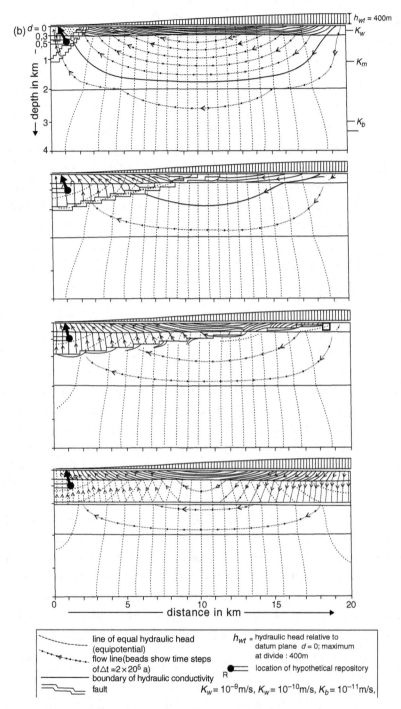

Fig. 5.49 (cont.)

'TAC continues to hold its previous opinion (TAC, 1989; TAC, 1990) that, from a groundwater hydraulics point of view, such regional recharge areas have positive characteristics as potential vault locations and therefore that a recharge area setting in regional groundwater flow systems should be formally recognized as a highly ranking favourable hydrogeological attribute' and a year later: 'we do record again our advice that in the development of the concept and its documentation, location in a regional recharge area be given serious consideration as a siting characteristic' (TAC, 1993, p. 7). The firm position of TAC was based on the present author's 'Recharge Area Concept' corroborated by numerical experiments (Tóth and Sheng, 1996). The experiments contrasted flow-path lengths and travel times of water particles issuing from repositories located in recharge and discharge areas at depths of 500 and 1000 m in hypothetical drainage basins representing generalized Canadian Shield conditions.

Figure 5.48 shows a computed flow net and isochrones of travel times for water particles infiltrating in the recharge area of a basin of given geometric and hydraulic parameters. The travel time of a water particle migrating along a flow line that crosses a repository is the difference between the particle's age when it leaves the repository and when it resurfaces at the flow line's discharge terminus (Fig. 5.48).

A significant observation is the relatively short time that the basin's oldest waters need to reach the surface in the vicinity of a repository in the discharge area. The observation dispels the myth that high age of groundwater indicates stagnant or quasi-stagnant flow conditions. The difference in flow-path lengths from the repositories to the land surface can be well appreciated visually (Figs. 5.49 a,b).

It is important to note also that in no case is contaminated water conveyed to the surface directly by the faults from the recharge-area repository regardless of their assumed attitude (Fig. 5.49a). On the other hand, flow up to the surface from the discharge-area vault is accelerated by additional recharge induced through the faults dipping down-slope (Fig. 5.49b). Figure 5.50 shows that travel times from the vaults to the surface are, on average, approximately one order of magnitude greater for repositories located in the recharge area than for those in the discharge area.

Whether it was the Technical Advisory Committee's insistence and/or arguments that had an effect is not known. Nevertheless, it appears from the Environmental Impact Statement that AECL's position has changed somewhat concerning the potential importance of recharge areas in finding suitable locations for repositories. To quote:

'In determining the technical suitability of a potential candidate area, the implementing organization would consider certain characteristics favourable, including: regional upland location, low topographic relief, few major lineaments, widely spaced major lineaments and few open fractures in the rock between lineaments (these characteristics would tend to maximize the time required for a contaminant

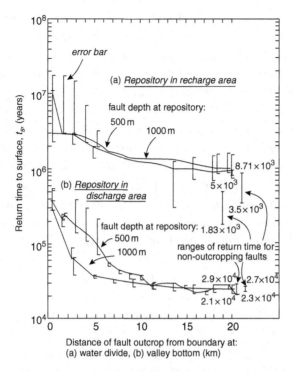

Fig. 5.50 Calculated return times of water to the land surface from points 500 and 1000 m below surface and 1 km from basin boundary in stratified basin with a fault: (a) repository in recharge area; (b) repository in discharge area (Tóth and Sheng, 1996, Fig. 9, p. 20).

released from a disposal vault to move through the geosphere to the surface.)' (AECL, 1994a, p. 154).

5.3.2 Sweden: (Swedish Nuclear Fuel Supply Co/Division Kärn–Bränsle–Säkerhet: SKBF/KBS)

The Swedish Parliament passed the 'Stipulation Act' in April 1977, concerning permission to charge nuclear reactors with fuel provided that the reactor's owner 'has demonstrated how and where an absolutely safe final storage of spent, unprocessed nuclear fuel can be effected' [SKBF/KBS–3 1983; I-General, p. 1:1; (Swedish Nuclear Fuel Supply Co/Division Nuclear Fuel Safety)]. In response, the Swedish nuclear power utilities organized the KBS (Kärn–Bränsle–Säkerhet: 'Nuclear Fuel Safety') project of the Swedish Nuclear Fuel Supply Co., for the purpose of investigating and reporting how the requirements set forth in the Act could be met. The potentially important role of groundwater flow was recognized early on in Sweden. 'A first investigation in the principal aspects of groundwater flow when varying topography, hydraulic conductivity and geometry' was presented by Stokes and Thunvik (1978, 'Summary'). The investigation was based on modelling ground-water flow in dozens of 2D 'synthetic' hydraulic sections according to the concepts attributed to Gustafsson (1970), namely: 'the groundwater level coincides with topography… The bottom part of the boundary is assumed impervious and is, in certain examples, of infinite depth. Finally, the area covered by the calculations is bounded by impervious vertical sides.' In addition, a number of hydraulic sections and travel times were calculated with different geometrical and hydraulic param-eters along actual transects in the vicinity of the town of Finnsjön. An example of the Finnsjön sections is reproduced in Figure 5.51, which clearly shows the effect of the topography and depth on the geometry and type of the induced flow systems (Stokes and Thunvik, 1978, Fig. 45, p. 55).

The clear-sighted conclusion, derived from the analysis and advocating a regional recharge position for repository location is best presented in the authors' own words (Stokes and Thunvik, 1978, p. 9) and is pictorially summarized in a later KBS publication (Fig. 5.52; SKBF/KBS−3, 1983; Fig. 5.52, III, Fig. 14-1, p. 14:3): 'Long flow times are obtained primarily beneath the inflow area (descending flow lines). One should then not select an area immediately beneath the highest point in the terrain since the outflow from this point is superficial and will reach the ground in a short time. Primarily one should study local terrain minima in large inflow areas. If the terrain minimum itself is an outflow area is of no significance. The local nature of the minimum causes up-currents only near the surface, while down-currents are obtained at greater depths. If the selected repository site is in this down-current, local outflow from the site is avoided. Here the determining factor is how deep the local flow extends, and this is determined by the depth-dependency of the hydraulic conductivity.' (See Figs. 3.3 and 3.11, respectively, for a method to estimate the depth of local flow systems, and for an illustration of the above described situation of flow in a local depression located in a regional recharge area.).

The merits of the 'recharge area concept' were underscored by the results of an advanced study of groundwater flow modelling conducted for the Swedish Nuclear Power Inspectorate (SKI) by Voss and Provost (2001) of the US Geological Survey. Although the results presented in their report are specific to the groundwater flow systems in the study area, the authors warn that they should only be considered 'as a demonstration of concepts and techniques for recharge-area siting of nuclear waste repositories in Sweden' (Voss and Provost, 2001, p.5).

The modelled area lies in southeastern Sweden. It stretches approximately 260 km north to south, 210 km east to west, covers approximately 49 000 km^2, and rises from

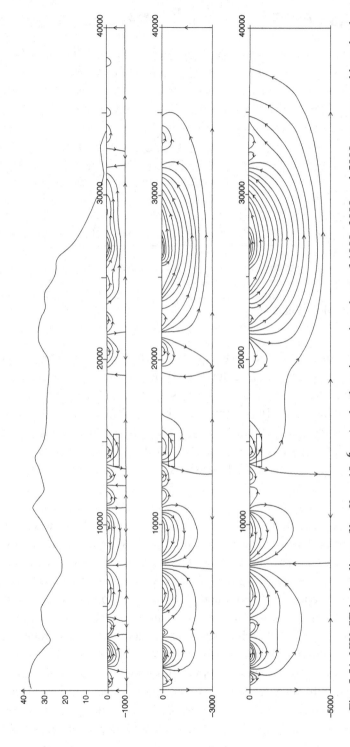

Fig. 5.51 NW–SE hydraulic profile: $K = 10^{-6}$ m/s; depth to impervious base of 1000, 3000 and 5000 m; water table at land surface; Finnsjön, Sweden (Stokes and Thunvik, 1978, Fig. 45, p. 55).

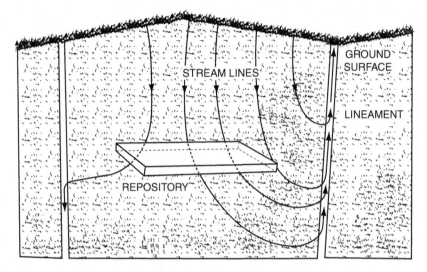

Fig. 5.52 Illustration of the principle of locating repositories in regional recharge area, recommended to the Swedish nuclear-waste isolation programme in 1983, and groundwater flow-pattern in the vicinity of a conductive fault (SKBF/KBS, 1983b, III-Barriers, Fig. 14-1, p. 14:3).

an elevation of 0 m at the Baltic sea to approximately 350 m at a distance of 123 km inland from the coast. The model extends vertically from the land surface to 10 km below sea level. The area was selected so as to include three potential candidate repository sites: Hultsfred, Simpewarp and Oskarshamn, located in discharge areas, as well as a hypothetical 'Comparison' site in a regional recharge location in the central highlands of the study area (Fig. 5.53). The US Geological Survey's SUTRA code was applied in three spatial dimensions to model variable-density flow and solute transport.

For each model, variant discharge areas were mapped and path length, travel time, and flow-path volume calculated (path length is the distance a water particle travels from a repository to the point where it reaches the water table; travel time is the duration a water particle needs to migrate from the repository to the water table; flow-path volume is 'the volume of rock encompassed by a radioactive plume emanating from a repository' (Voss and Provost, 2001, p. 3). The concept of the flow-path volume has been introduced by Voss and Provost (2001). It is an additional safety factor since larger rock volumes provide greater potential for retardation (by adsorption and dilution) of radio nuclides.

A section of computed return-flow times illustrates 'how the undulating topography in southeastern Sweden may generate both regional and nested local flow systems' (Fig. 5.54; Voss and Provost, 2001, p. 13). Repository locations within the dark grey band have the longest return-flow times. The primary regional system

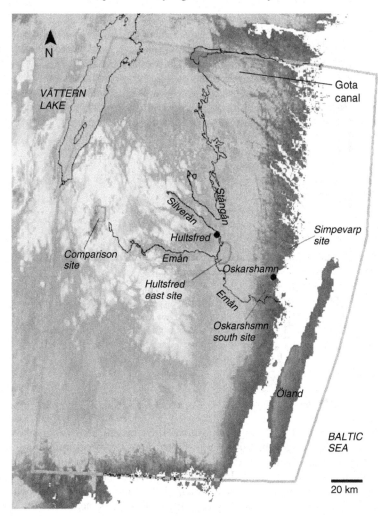

Fig. 5.53 Area of groundwater of flow-model showing topography, potential nuclear fuel-waste repository sites, and the 'Comparison' site, southeastern Sweden (Voss and Provost, 2001, Fig. 4).

in this section has recharge below the topographic peak and discharges both to the left boundary (near Vättern Lake) and to the Baltic Sea (top right boundary) The 'Comparison' site is located below the primary recharge area. Consequently, water starting its underground journey there takes the longest time to return to the surface. The Hultsfred site spans both low (on the west) and very long return-flow times (on the east). The coastal sites Simpevarp and Oskarshamn, on the other hand, are in a major discharge area and, consequently, have relatively short return-flow times.

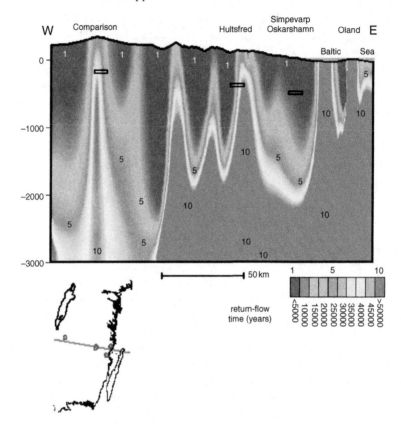

Fig. 5.54 Cross-section through repository sites in the area of groundwater flow model (Fig. 5.53) showing return-flow times for water migrating from repositories located at any specified position in the flow domain to the water table; an illustrative simulation. Inset shows section location (heavy black line) and repository sites 'Comparison', Hultsfred, Simpervarp, and Oskrshamn in Figure 5.53, southeastern Sweden (modified after Voss and Provost, 2001, Fig. 3).

In summarizing the study's findings, Voss and Provost (2001, p. 53.) state:

'In order to improve chances that a repository site will maximize the safety-related factors, path length, travel time and path volume, the following siting objectives need to be met:

- The repository should be located below a ground-water recharge area.
- The repository should be located far from potential discharge areas for the site.
- The repository should be located in a contiguous region of high return-flow time, large enough to include the repository.
- The repository should be located in region where ground-water flow is downward for considerable (a few kilometers) depth below the repository.'

5.3.3 Switzerland: Nagra (Nationale Genossenschaft für die Lagerung radioaktiver Abfälle)

Upon enactment of the 'Atomgesetz' (Atom Law) in October 1978, the Swiss 'Nationale Genossenschaft für die Lagerung Radioaktive Abfälle' (National Corporation for Nuclear Fuel-Waste Storage) was commissioned by the country's nuclear power-plant operators to undertake 'Projekt Gewähr 1985' ('Project Guarantee 1985') with the objective to show by the end of 1985 that lasting and secure handling and storage of spent nuclear fuel can be guaranteed. A report titled '*Projekt Gewähr 1985*' was completed by 23 January 1985, and its evaluation by the relevant Swiss authority, Hauptabteilung für die Sicherheit der Kernanlagen (Swiss Federal Nuclear Safety Inspectorate), was summarized in HSK (1987).

The study focused on a site of a hypothetically planned repository in the Rhine valley in northern Switzerland, and hydrogeology was a major aspect of the investigations. In addition to the intensive drilling, testing, and sampling activities at and in the vicinity of the site, two 3D mathematical models of groundwater-flow were constructed: one 'regional' model covering an area of approximately 23 000 km^2, including a large portion of the northern Alps, parts of the Swiss Jura mountains and reaching to the Schwarzwald (Black Forest) on the north; and one 'local' model, centred on the hypothetical repository site and its closer surroundings and covering 900 km^2 (Kimmeier *et al.*, 1985). One of the most significant conclusions derived from the numerous flow-pattern variants calculated for the different sets of geologic configurations and hydraulic parameters was that the hypothetical repository site was in an area of concentrated regional groundwater discharge (Fig. 5.55; Kimmeier *et al.*, 1985, Annexe 8-2). This conclusion was corroborated by the results of field investigations and led to the summary opinion of the official expert review (HSK, 1987, p. 5-46): 'Die bisherigen Modellierungen sowie die Ergebnisse der Nagra Sondierbohrungen zeigen, dass das postulierte Endlagergebiet generell im bereich einer Exfiltrationszone der kristallinen Tiefenwässer liegt. Diese Situation ist grundsätzlich ungünstig.' ('The modelling to date, as well as the data from Nagra's test holes show that the hypothetical repository site is located in a general discharge area of groundwater from the deep crystalline basement. This situation is fundamentally unfavourable.' Translation by the author.)

5.3.4 U.S.A: Palo Duro Basin, Texas

One of the geologic formations that the US Department of Energy considered in the early 1980s as possible host rock for a future national repository of nuclear-fuel waste was the Evaporite Aquitard of Permian age in the Palo Duro Basin of the Texas Panhandle, USA. Two major groundwater-related component studies of the

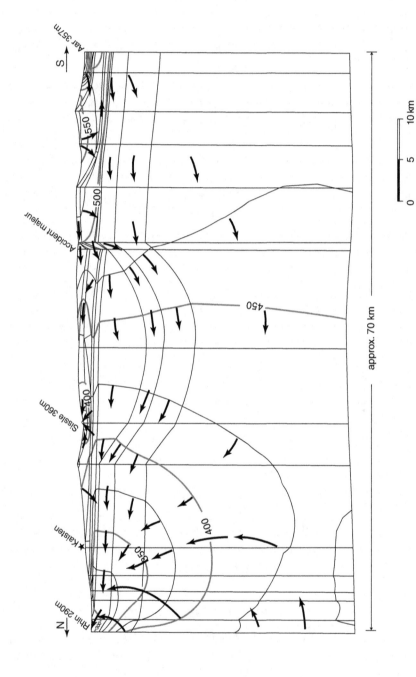

Fig. 5.55 Calculated groundwater flow pattern along section from the Aare River in the Alps, to the area of concentrated discharge in the Rhine valley on the north. Potentiometric values in m (Kimmeier et al., 1985, Annexe 8-2).

investigations were: (i) a field-data based reconstruction of the area's subsurface flow distribution (Bair and O'Donnell, 1985; Bair, 1987) (ii) a theoretical evaluation of the temporal evolution of groundwater flow patterns during post-Miocene times by numerical modelling (Senger *et al.*, 1987; see also Ch. 3.3.2, Fig. 3.27).

Bair and O'Donnell (1985) identified three regional-scale hydrostratigraphic units (HSU), namely, HSU A, HSU B and HSU C, in descending order (Fig. 5.56).

With the land surface being its top, HSU A is a 245–335 m (800–1100 ft) thick unit consisting of alluvial, fluvial and lacustrine deposits of the Tertiary Ogallala Formation and the Triassic Dockum Group. 'Even though confined, semiconfined, and unconfined flow conditions exist in the Ogallala/Dockum, on a regional scale HSU A comprises a shallow, usually freshwater system' (Bair and O'Donnell, 1985, p. 2).

HSU B is 1250–1430 m (4100–4700 ft) thick and made up of several formations of shales, siltstones, evaporites with some carbonates and sandstones. It is characterized by extremely low permeability and, on a regional scale, acts as a major confining unit between HSU A and the underlying aquifer system in the deep-basin strata. On a local scale some strata in HSU B are sufficiently porous and permeable to store and transmit fluids, which include oil and gas in the area.

HSU C is 850–1340 m (2800–4400 ft) thick and consists predominantly of the Lower Permian Wolfcamp Series and the Pennsylvanian System, but also includes strata from Mississipian through Cambrian age.

'A major purpose of the hydrodynamic investigation is to determine the vertical direction of flow in the shale and evaporite aquitard and the host rock that underlie the major freshwater aquifer system and overlie a deep-basin brine aquifer system' (Bair and O'Donnell, 1985, p. 2).

The subsurface flow distribution was portrayed by potentiometric surface maps, hydraulic cross-sections, pressure-depth diagrams, and maps of hydraulic-head difference. The data for the construction of the maps, sections, and pressure profiles were culled from thousands of water levels from inventoried wells, thousands of drill-stem tests (DST), and twenty long-term pumping tests.

Figure 5.56 shows a southwest-northeast hydraulic section across the study area. Quoting freely from Bair and O'Donnell (1985, p. 7-8, Figs. 2 and 11; Bair, 1987, Figs. 8B, 9B), the section was constructed using water-level data from the Ogallala and Dockum aquifers and from equivalent freshwater hydraulic heads calculated from DSTs and long-term pumping tests performed in HSU B and HSU C. The figure shows that most of the water in HSU A flows horizontally and discharges to springs and wells. Nevertheless, a potential gradient exists for some water to flow downward from HSU A across HSU B, and into HSU C. It also shows a local flow system in the northeastern part of the profile where water is ascending in the upper part of HSU B. This observation explains the near-surface salt-dissolution

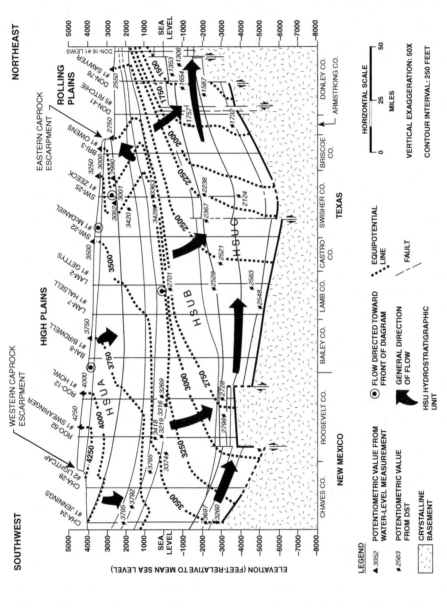

Fig. 5.56 Southwest–northeast section of Hydrostratigraphic Units HSU A, HSU B and HSU C, and potentiometric profile, based on water-level measurements in HSU A and equivalent freshwater hydraulic heads from DSTs in HSU B and HSU C, Palo Duro Basin, Texas, USA (modified from Bair and O'Donnell, 1987, Figs. 2, 11; Bair, 1987, Figs. 8B, 9B).

zones in areas adjacent to the Eastern Caprock Escarpment (Gustavson *et al.*, 1980; Boyd and Murphy, 1984) and the brine-emission areas in the Rolling Plains region (Engineering Enterprises, Inc., 1974). The discharging brines are derived from shallow meteoric origin rather than from a deep basinal origin (Kreitler and Bassett, 1983). Figure 5.56 shows also that water in HSU C is a mixture derived from two sources: (i) recharge from the outcrop area of the deep-basin strata in east-central New Mexico; (ii) leakage across HSU B.

In the modelling part of the investigation, Senger *et al.* (1987) used a two-dimensional groundwater flow model along the same southwest–northeast oriented and approximately 535 km long cross-section through the basin 'to characterize regional ground-water flow paths as well as to investigate underpressuring below the evaporite aquitard, to evaluate mechanisms of recharge and discharge to and from the deep-basin aquifer, and to examine transient effects of erosion and hydrocarbon production. This study was designed to investigate various factors affecting the overall ground-water flow pattern in the basin and was not necessarily aimed at producing a fully calibrated predictive model' (Senger *et al.*, 1987, p. 3). In the first part of the two-phase study the objectives were 'to evaluate the effects of hydrostratigraphy and topography on the regional hydrodynamics of the basin, with special emphasis on explaining the observed regional-scale underpressuring of the deep brine aquifers. In the second phase, long term transient flow conditions caused by different tectonic and geomorphologic processes were investigated' (Senger *et al.*, 1987, p. 3–4).

Figure 3.27 (Senger *et al.*, 1987, Fig. 12, p. 19) illustrates one of the several dozen model variants and shows: *local* and *intermediate* flow systems (terms applied at the given, regional scale!) induced by the Pecos River Valley in New Mexico on the west; a regional system recharging through the surface of the high plains at the Texas–New Mexico boundary and moving above the evaporite aquitard to the discharge area at the eastern Caprock escarpment; a second, basin-wide system from the west end of the section, stretching beneath the evaporite aquitard through the granite wash beds of the deep basin brine aquifer and leaving the section through the eastern boundary; and significant underpressuring in the latter aquifer with respect to the hydraulic heads in the shallower zones of the Texas high plains.

Underpressuring in the deep basin brine aquifer would be a favourable attribute of a repository located in the evaporite aquitard above. Possibly escaping radionuclides would be forced downwards, away from the surface, and into the long and sluggish easterly moving regional flow system, one expects they would have time to lose their radioactivity before returning to the biosphere.

The underpressuring of the deep basin brine aquifer was attributed to two principal factors: small inflow into its regional flow system from the west, and high permeability of its granite wash members, thus facilitating a high rate of discharge on the east to the low lying plains in Oklahoma. The vertical leakage through the evaporite aquitard into the deep basin aquifer is relatively small at 0.0125 m³/day, due primarily to the low vertical hydraulic conductivity of the aquitard's compo-nent strata which vary between 10^{-9} and 10^{-12} m/s. The low rate of recharge from above and the relatively high rate of lateral discharge to the east, owing to the high hydraulic conductivity of granite wash ($\approx 10^{-6}$–10^{-7} m/s), combined with the low-altitude discharge areas create the interesting 'tensiometric effect', discussed in Section 3.2, Figure 3.11.

5.4 Interpretation and utilization of observed deviations from theoretical patterns of gravity-driven groundwater flow

Subsurface flow-patterns inferred from observed well-water levels, pore pressures, water chemical compositions, and field manifestations of groundwater are fre-quently different from calculated flow fields based on assumed properties of the flow domain such as boundary conditions and hydraulic parameters. The most common possible reasons for discrepancies are actual differences between the real-life and the assumed conditions in: (i) the source of the water-driving force; (ii) the bound-ary conditions of the force field; (iii) the values and distribution patterns of the rock framework's permeability; (iv) the state of flow, namely, steady or transient.

Deviations in themselves are no reason to discredit the model. In fact, the differ-ences between calculated and expected conditions can be used to advantage in fine tuning the model to improve its use for its intended application. They may even lead to discoveries of unsuspected properties of the flow field or the rock framework, or both. Such properties may be valuable in various ways. Indeed, the principles, processes, and purpose of deliberately searching for 'fluid-potential anomalies', are comparable to those of geophysical exploration based on anomalies of the Earth's gravitational, magnetic or electrotelluric fields. The examples below, most of which are taken from the author's personal experience, are intended to point the way in this direction.

5.4.1 Highly permeable rock pods

Rock bodies of given permeabilities emplaced in a rock matrix of contrasting per-meability modify the flow field as compared to one embedded in similar rock. The degree of modification depends on the rock body's size, shape and permeability relative to the surrounding matrix. Similar to the effects of a soft iron core or an antimagnetic body placed in a homogeneous magnetic field, highly and poorly per-meable rock bodies cause the flow lines to converge through, or to bend around, them, respectively. A modification of the flow lines requires a modification of the equipotential lines. The difference between the original and the modified fluid-potential values is a potentiometric anomaly. It can be calculated if the rock's hydraulic properties are known (Section 3.2.2, Fig. 3.15). The theory of 'hydraulic exploration' by rock-pod generated fluid-potential anomalies is based on the flowing water's transportation ability and on the capacity of a relatively highly permeable rock body to retain and store fluid.

Based on the above considerations, Tóth and Rakhit (1988) used 'hydraulic exploration' to locate and outline rock-bodies of petroleum-reservoir quality in the Lower Cretaceous Bow Island Formation at Keho Lake in southern Alberta, Canada.

First, a local area of potentiometric perturbations, or anomalies, was selected from a larger map reduced from drill-stem-test (DST) measurements of formation-pressure (Fig. 5.57). At the same time, interval-permeability values were calculated from analyses of drill cores, and cross-sections and a map of their spatial distribution was prepared [interval-permeability = permeability k (millidarcy)×formation thickness d (metre)]. This distribution was to be simulated iteratively by 'inverse modelling'.

Next, a succession of 'iterative' potentiometric maps was calculated (e.g., Fig. 5.58). Constant-head boundaries were based for these maps on the head values observed several kilometres outside the local study area. The boundary values were kept constant, while the values and spatial distribution of the interval permeabilities were varied by discrete steps in attempting to improve the match with the mapped potentiometric surface (Fig. 5.57; inverse modelling). Figure 5.59 shows the areal distribution of the computed interval permeabilities resulting in the best potentiometric match. The section of structure and interval permeability through lenses L_4, L_1 and L_2 (Figs. 5.59 and 5.60) confirms the validity of the computed results.

One possible incentive to explore for highly permeable rock bodies is that they can capture and retain oil and/or gas and thus become commercially valuable petroleum reservoirs. Owing to the grain-size difference between a sandstone lens and its surrounding shale matrix, a capillary barrier develops at the lens's boundary. A volume element of petroleum straddling such a boundary in a hydrostatic environment has the tendency to be pushed by capillary forces from the fine- into the coarse-grained rock and to resist movement in the opposite direction (Hubbert, 1953, p. 1975–1979). These tendencies are enhanced and reduced by formation-water flow if the flow is in the direction of fine- to coarse- or, respectively, coarse- to fine-grained rock (Fig. 5.61). The resulting process of gradual increase of petroleum saturation with time inside the lens has been demonstrated by a numerical experiment and is illustrated in Figure 5.62 (Rostron and Tóth, 1989, Fig. 6, p. 45).

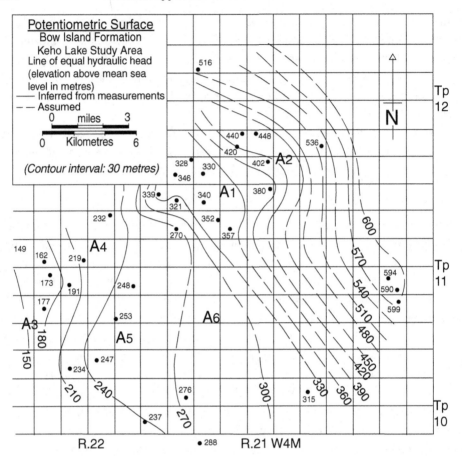

Fig. 5.57 Potentiometric surface map inferred from DST-measurements of pore pressures, Bow Island Formation, Keho Lake study area, S. Alberta, Canada (Tóth and Rakhit, 1988, Fig. 13, p. 372).

5.4.2 *Hydraulic barriers to flow*

Hydraulic barriers, or negative hydraulic boundaries, such as slightly permeable fault plains, un-fractured igneous dykes, lithologic pinch-outs, argillaceous walls of buried river valleys, and so on, can be a boon to the hydrogeologist, as in containing contaminant migration, or a bane, when reducing source areas to water supply wells. It is desirable, therefore, to be aware of such subsurface features prior to the design of a groundwater project that is possibly affected. Analyses of the undisturbed natural potentiometric patterns can be helpful in detecting them. The exploration programme for the Town of Olds, Alberta, Canada, is a case in point (Section 5.2.1, Figs. 5.12, 5.14, 5.15; Tóth, 1966b).

The potentiometric surface shown in Figure 5.63 was constructed from water levels obtained from the area's farm wells and the project's test holes prior to

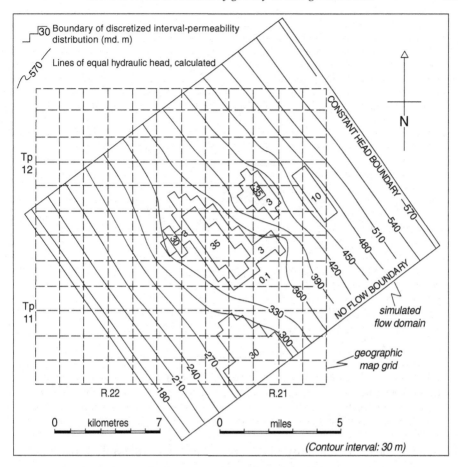

Fig. 5.58 Final simulation of iterative inverse modelling of potentiometric surface based on measured values of permeability and constant head boundaries inferred from field data (Tóth and Rakhit, 1988, Fig. 18, p. 375).

designing the field of three production wells that subsequently would supply the Town of Olds. Some puzzling features of the potentiometric configuration were: the abrupt southern termination of the oval shaped flowing-well region lying north east of the study area's centre; the sharp drop of hydraulic heads just south of it (Figs. 5.12, 5.15, 5.63); and the southwards protruding equipotential lobes leading from the north into the oval artesian region (Fig. 5.63).

Just south of the 'potentiometric step' the undisturbed water-level is uncommonly flat but it has a well developed spur pointing straight west. Because of the extremely heterogeneous character of the Upper Cretaceous floodplain deposits of the Edmonton Formation, no discrete lithologic units could be identified. Consequently, the hypothesis could not be verified by geological means according to which the above phenomena collectively are caused by an east–west striking

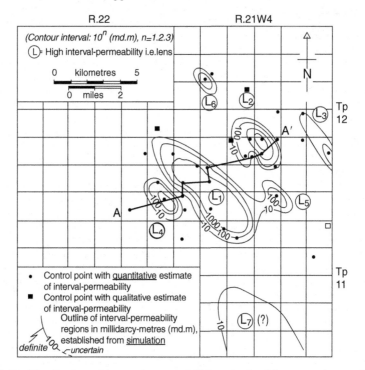

Fig. 5.59 Distribution of interval permeabilities obtained by step-wise modification of observed field values until optimum agreement between observed (Fig. 5.57) and simulated (Fig. 5.58) potentiometric surfaces was reached (modified after Tóth and Rakhit, 1988, Fig. 19, p. 375).

subvertical linear discontinuity in the rock framework, possibly a fault. Instead, two pumping tests of several days duration were conducted, both north and south of the suspected lithologic discontinuity. Figure 5.64(a) and (b) show the evolution in time of the induced potentiometric depressions. The configuration of the draw-down pattern proves the presence of an approximately east–west oriented hydraulic barrier between the two pumped wells. The northwest extending draw-down trough induced by test 2 in the northerly well (Fig. 5.15, Olds Well No. 189; Fig. 5.64a) explains the southward protruding equipotential lobe of the pre-pumping situation. And the long and deep east–west striking hydraulic depression due to test 3 in Olds Well No. 192 south of the barrier (Figs. 5.14, 5.64b) confirms the presence of an elongated high-permeability rock pod: a sand and gravel fill of a paleo-river bed indicated earlier by the natural potentiometric pattern (Fig. 5.14).

5.4.3 Abrupt change in chemistry across flow-system boundary

Abrupt lateral and/or vertical changes in the chemical composition of groundwater are common. In many instances such changes are clearly linked to lithologic changes

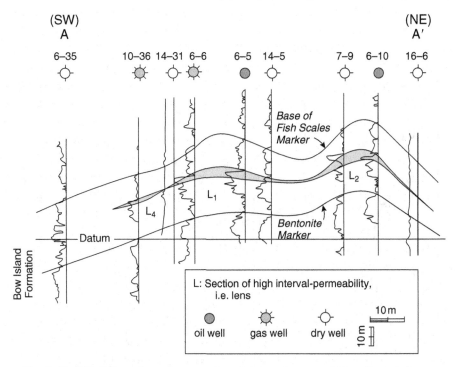

Fig. 5.60 SW–NE section of structure and interval-permeability through lenses L_4, L_1, and L_2 (see Fig. 5.59; Tóth and Rakhit, 1988, Fig. 16, p. 374).

or discrete proximate aquifers. However, in rock frameworks without distinct litho-logic or structural contrasts, an explanation may be found in the distribution pattern of groundwater flow systems.

Figures 3.2 and 4.1 illustrate the conceptual basis of this argument in vertical sections: several flow systems that have a common area of origin can split and terminate in widely separate locations. Or, the other way around, flow systems originating in distant regions of different geologic makeup can merge into a com-mon discharge area. In both cases, along certain segments of flow systems water can migrate next to water with a differing hydrogeochemical history and thus a differing chemical composition. Similarly, abrupt changes in water-chemical types can also occur laterally, across vertical boundaries between adjacent flow-systems that traverse different portions or different lengths of the rock framework.

In an exploration project for the municipal water supply of the Town of Three Hills, central Alberta, Canada, a change from 1500 to 2500 mg/l and more in total dissolved solids content TDS was noticed over a distance of less than 500 m along Section H_2–H_2' (Fig. 5.65). In the unstratified and 'poorly sorted continental, argillaceous, bentonitic, crossbedded, lenticular, slightly indurated siltstones, and

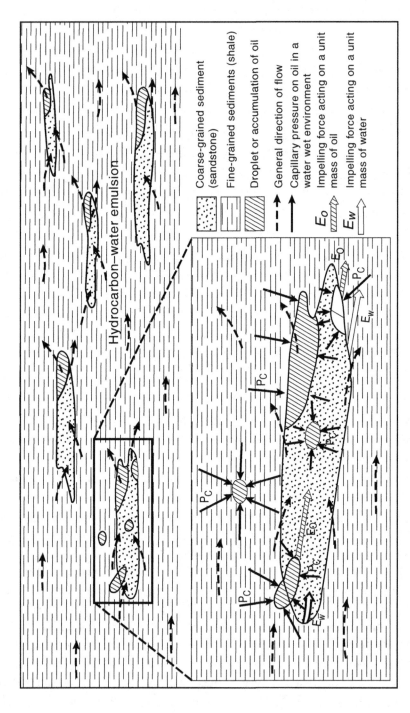

Fig. 5.61 Interplay of capillary and impelling forces acting upon volume elements of hydrocarbons at sand–shale interfaces and possibly resulting in entrapment at the 'downstream' end of sandstone lenses (Tóth, 1988, Fig. 3, p. 487).

2 000 000 YEARS

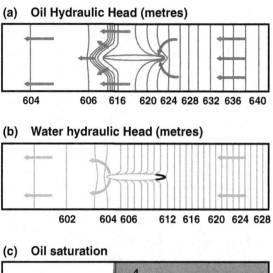

(a) Oil Hydraulic Head (metres)

604   606   616   620 624 628 632 636 640

(b) Water hydraulic Head (metres)

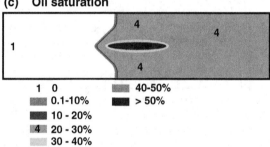

602   604 606   612 616 620 624 628

(c) Oil saturation

1 0	▨ 40-50%
▨ 0.1-10%	■ > 50%
■ 10 - 20%	
4 20 - 30%	
▨ 30 - 40%	

Fig. 5.62 Modelled distributions of hydraulic heads for oil and water and of oil saturation in high-permeability sandstone lens enclosed in a shale matrix 2×10^6 years after import of oil by water flow began at right end of the flow domain (Rostron and Tóth, 1989, Fig. 6, p. 45).

sandstones, coal, volcanic tuff and clay ironstone' rocks of the Upper Cretaceous and Paleocene Paskapoo and Edmonton Formations neither lithology nor hydrostratigraphy could explain the abrupt change in the water's chemical type (Tóth, 1968, Fig. 5, p. 14).

The highest TDS values in the area were found in a tract of groundwater discharge, marked by flowing wells and a north–south running potentiometric trough (Fig. 5.66). To the south–east of the TDS maxima, an area of flat potentiometric surface extended 3–4 km towards the 'Three Hills Ridge'.

An explanation for the sharp change in groundwater chemistry was sought in flow system analysis. To this end, conductive-paper electric analogue models were constructed along some west–east cross-sections based on observed well-water

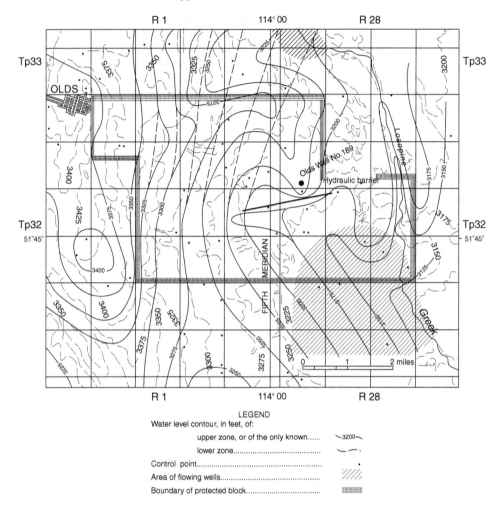

Fig. 5.63 Non-pumping groundwater levels prior to large scale development for municipal water supply, Olds groundwater exploration area, Alberta, Canada (after Tóth, 1966b, Fig. 9).

levels and the topography (Fig. 5.66). Section H_2-H_2', running along the W–E centre line of the map in Figures 5.65 and 5.66 provided the key to the puzzle (Fig. 5.67).

The maximum TDS values occur approximately 3.2 km (2 miles) east of the map's western boundary. This region coincides closely with the center of the contiguous discharge areas of two regional flow systems that converge from west and east and merge here (between miles 7 and 8 on Fig. 5.67). On the other hand, the 1500 mg/L TDS contour runs within 200–300 m east of the modelled boundary between the discharge area of the regional flow system sourced in the Three Hills

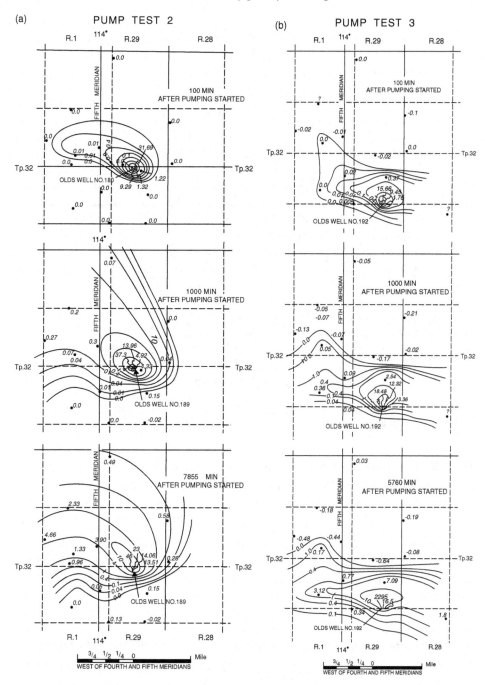

Fig. 5.64 Changes in groundwater levels in response to pumping in: (a) Olds Well No. 189, (b) Olds Well. no. 192, Olds groundwater exploration area, Alberta, Canada (after Tóth, 1966b, Figs. 18, p. 43, and 19, p. 44).

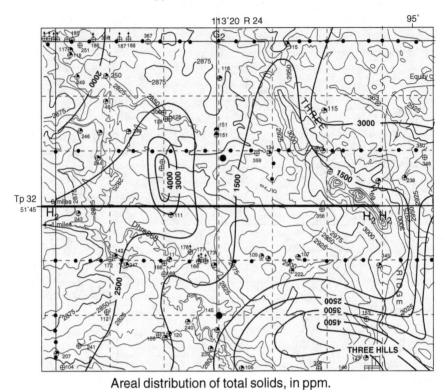

Areal distribution of total solids, in ppm.

Fig. 5.65 Groundwater mineralization, TDS, in ppm = mg/l; grid spacing: 1 mile = 1.6 km (Tóth, 1968, detail of Fig. 17, p. 43).

Ridge, 5 km (≈3 miles) farther east, and the discharge areas of shallow (100–300 m) and short (800–1500 m) local systems. The solution to the question was that the discharge area of the long, thus relatively highly saline, regional system is adjacent to the discharge area of the less saline waters of the short local systems (Fig. 5.67). A note of historical interest: numerical groundwater modelling was still in its infancy during the early 1960s, and the above conductive-paper analogue model was reviewed both by Domenico (1972, Fig. 6.8, p. 268) and Freeze and Cherry (1979, Fig 5.11, p. 180).

5.4.4 Identifying mechanisms of subhydrostatic pore-pressure generation

Subhydrostatic pore pressures, or underpressuring, can be induced by various geologic mechanisms. Some of the most common and most effective of these are the draining effect of low-altitude aquifer outcrops, elastic rebound of the rock framework due to erosional or/and glacial unloading, and delayed adjustment to increasing pore pressure due to sediment loading during subsidence. In all cases,

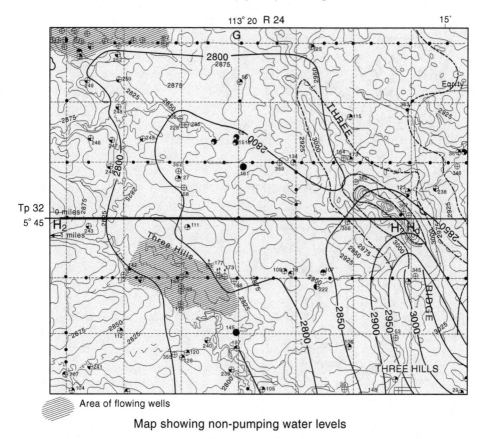

Map showing non-pumping water levels

Fig. 5.66 Non-pumping groundwater levels, contour values in feet; grid spacing:
1 mile = 1.6 km (detail of Fig. 10, p. 23, Tóth, 1968).

the presence of low-permeability hydrostratigraphic units tend to magnify the
effect. The actual cause of underpressuring may be determined in certain cases
by comparing the *real pattern* of subsurface fluid-dynamic parameters (inferred
from observations) with *reference patterns* (generated for rigorously quantifiable
conceptual conditions).

A theoretical, or reference, case for the effect of low-altitude outcrops, formulated
originally to explain subhydrostatic pressures in the Red Earth region of northern
Alberta, Canada (Tóth, 1978) is reproduced in Figure 3.12. The salient features
of this pressure pattern in the context of underpressuring are: the sharp drop in
the pressures (not just in the pressure gradient) beneath the aquitard in the regional
recharge area (sites 2 and 3); hydrostatic values of the vertical gradients in the region
of horizontal flow (site 1); and slightly superhydrostatic pressures and pressure
gradients, indicating an upward flow component in the regional discharge area
(site 4).

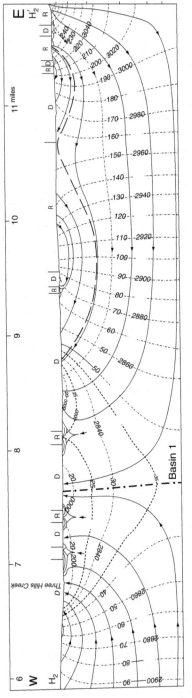

LEGEND

2900 ········· Line of equal hydraulic head, feet amsl

40 ········· Line of equal electric potential above reference value, mV

—————— Line of fluid driving force

— — — — Boundary between flow systems

—··—··— Boundary between groundwater basins

|R| Recharge area (downward flow)

|D| Discharge area (upward flow)

Vertical exaggeration $\dfrac{M_v}{M_h} = 2$

Fig. 5.67 Part of electric analogue Section $H_2–H_2'$, contour values in feet; distance scale along top of section in miles, 1 mile = 1.6 km (Tóth, 1968, detail of Fig. 13).

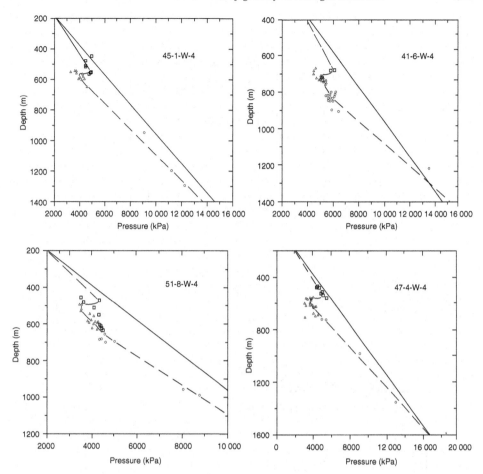

Fig. 5.68 Underpressuring due to low outcrop elevation of the Manville aquifer enhanced by the overlying Lea Park–Colorado shale aquitards, Chauvin area, east-central Alberta, Canada (Holysh and Tóth, 1996, Fig. 10, p. 265). Key: dashed lines = observed data; solid lines = hydrostatic gradient; point symbols: measured pressures in aquifers (reprinted by permission of the AAPG whose permission is required for further use).

The similarity between the pressure-depth profiles in the Chauvin area of east-central Alberta, Canada (Fig. 5.68; Holysh and Tóth, 1996) and those in Figure 3.12b, was interpreted as an indication of the regional pressure-drawdown effect of the Mannville aquifer subcropping at lower elevations tens of kilometres farther east in the province of Saskatchewan. The conclusion was supported by hydraulic cross-sections showing descending water movement (Fig. 5.69), and by potentiometric maps (not shown here) indicating easterly flow in the Mannville Group.

Figure 5.70 shows measured hydraulic heads (shown as converted to pressures) declining monotonously with depth toward the boundary value of 1 atm in a tunnel,

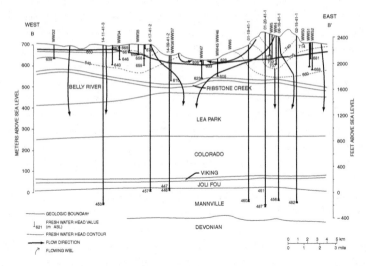

Fig. 5.69 E–W hydrogeologic section showing local flow system and downward flow through the Lea Park–Colorado shale aquitard into the Mannville aquifer, Chauvin area, east-central Alberta, Canada (Holysh and Tóth, 1996, Fig. 11b, p. 267; reprinted by permission of the AAPG whose permission is required for further use).

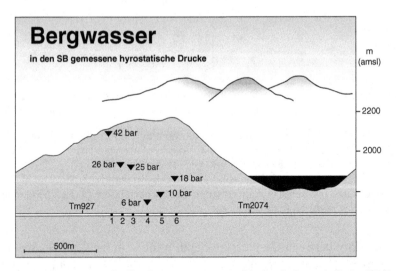

Fig. 5.70 Monotonous decline in pore pressure with depth through Swiss Wellenberg mountain toward atmospheric value in tunnel ≈450 m below mountain top (diagram by J. Tóth).

through an approximately 450 m thick section of a mountain in Switzerland. In addition to illustrating the effects on flow patterns of low altitude discharge boundaries, the case evidences bulk hydraulic continuity in granitic rocks by the abundant leakage from the tunnel's walls (Fig. 5.71).

On the basis of conceptual arguments Tóth and Corbet (1986) proposed that certain petroleum deposits in the Taber area, southern Alberta, Canada, were emplaced

Fig. 5.71 Water seepage into tunnel through more than 400 m of granitic rock in Swiss mountain (photo by J. Tóth).

by gravity-driven groundwater flow systems during the period of 35–5 Ma in mid-to-late Miocene times. However, the argument went, the gravity-flow patterns must have been obliterated by the currently existing strong underpressuring. The underpressuring itself was attributed to elastic rebound and dilation of the low-permeability Colorado shale aquitard upon erosional removal of more than 700 m of overburden during the last 10–20 million years (Figs. 5.72 and 5.73).

The theory was tested subsequently by Corbet and Bethke (1992) by quantitatively modelling the effects on pore pressures of the arguably most important modifying factors, namely: elastic pore dilation due to unloading by overburden removal; thermal contraction of cooling pore–fluid volumes due to reduced depth of the investigated rock units beneath the eroding land surface; and the extant part of the original hydrostatic pressure remaining after periods of erosion, due to the time lag in pressure dissipation behind the rate of land-surface lowering, called 'memory' by Corbet and Bethke (1992). Using realistic formation parameters and geologic

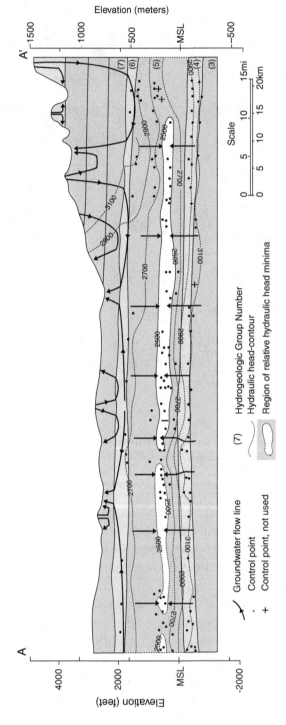

Fig. 5.72 Present hydraulic-head distribution and inferred formation-water flow pattern based on observed pore pressures, Section A–A', Taber area, southern Alberta, Canada (after Tóth and Corbet, 1986, Fig. 24, p. 357).

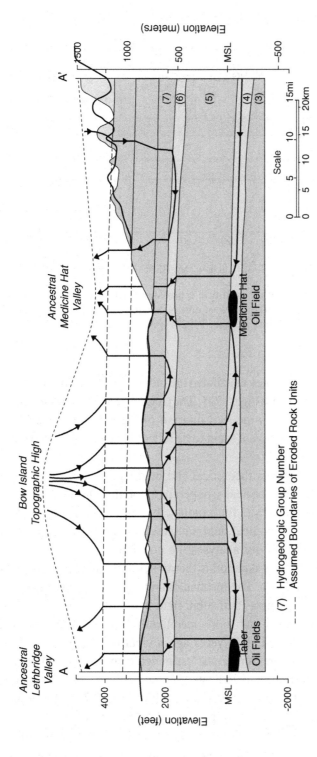

Fig. 5.73 Reconstructed mid-to-late Miocene topography hypothetically generating the formation-water flow pattern which, in turn, may have resulted in the known petroleum accumulations, Section A–A', Taber area, southern Alberta, Canada (after Tóth and Corbet, 1986, Fig. 28, p. 361).

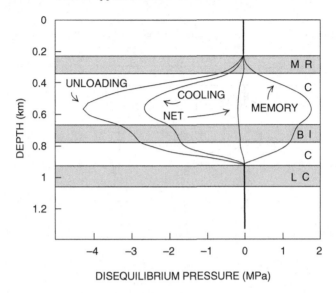

Fig. 5.74 Components of disequilibrium pressure for the present day calculated for a vertical profile near the the west end of Section A–A′, Taber area, southern Alberta, Canada (Corbet and Bethke, 1992, Fig. 13, p. 7212).

time estimates for the area, they calculated the individual effects of the above factors and the resulting 'net' sum (Fig. 5.74). They concluded that the underpressuring can be explained by erosion-induced dilation of the rock framework. However, paleo-topographic effects did not seem to be reflected in the calculated current flow pattern.

Vinard (1998) conducted a theoretical and field-based study of underpressured conditions for the Swiss National Cooperative for the Disposal of Radioactive Waste (Nagra) at a potential site for a radioactive waste repository at Wellenberg, Switzerland. Based on geological field data, dedicated hydrogeological tests, and extensive numerical sensitivity analyses, he concluded that: 'it can be affirmed that deformation of the rock skeleton is the most important mechanism leading to subnormal pressure' and that 'The mechanical rebound of the marl-shale aquitard, which led to the currently measured subnormal hydraulic heads, is predominantly generated by glacial retreat and more specifically by glacial melting at the end of the last glaciation some 12 000 years ago' (Vinard, 1998, p. 198). The timing and duration of the underpressuring was estimated from the rates of pore-pressure changes. The cause, however, is indicated by the geometry of the underpressured zone and the shape of the vertical pressure profiles (Fig. 5.75). Figure 5.75 shows one of the numerous simulations suggesting dilation of the rock as being the plausible cause of underpressuring by (a) the great lateral extent of the affected rock volume and (b) the monotonously decreasing hydraulic heads towards the centre of the dilated rock mass from hydrostatic values both at its top and at its bottom (cf. Fig. 5.72).

In a study in west central Alberta, Canada, Parks and Tóth (1995) found a zone of maximum underpressures between 500 and 1000 m below the present land surface in Upper Cretaceous and Tertiary rocks (Fig. 5.76). Vertical pressure gradients in overlying and underlying strata are interpreted to indicate vertical fluid-driving forces convergent into this zone. Lateral gradients indicate fluid flow towards the area in which maximum erosional stripping has occurred since mid Tertiary (Price, 1994). Rock-pore dilation in hydraulically 'tight' strata due to erosional unloading is the best explanation for these observations.

5.5 Exploration for petroleum and metallic minerals

A theoretically justified but in practice still underused application of groundwater flow studies, in general, and gravity-flow studies, in particular, is for exploration for petroleum and some metallic minerals. The conceptual basis of the approach is straightforward: petroleum and many metallic minerals are aggregated into commercially significant deposits by moving groundwater at locations that can be anticipated based on fluid dynamic, physical, chemical, and/or environmental considerations. They may also be ablated and carried away from such deposits leaving interpretable traces behind. Knowledge of past or current groundwater flow patterns may thus shed light on expected sites of accumulation either by pointing forward to possible destinations of migrating mineral constituents, or backward to possible sources of origin, or both.

Explorationists have long since been aware of the fact that many minerals migrate by aqueous transport. They have, however, failed to consider the key factor that determines the flow paths, namely, the spatial distribution of fluid potential. The basic thinking in petroleum exploration illustrates the point. The two principal geologic parameters used by the traditional petroleum geologist are the attitude, or dip, of permeable avenues of the rock framework (strata, faults), and sedimentary compaction, as the commonly assumed source of abnormally high formation pressures. Because sedimentary compaction is thought to be maximum in the centre of depositional basins, subsurface fluids are commonly believed to migrate up-dip, i.e. centrifugally out from the depocentres. Down-dip flow, i.e. centripetal flow directions, due to gravity flow driven seaward from elevated terrestrial areas and its possible consequences have not normally been taken into account.

It is only natural that a new concept and its practical implementation should have started with minerals that are readily amenable to aqueous transport in the subsurface. Consequently, the currently available examples also are limited to potential

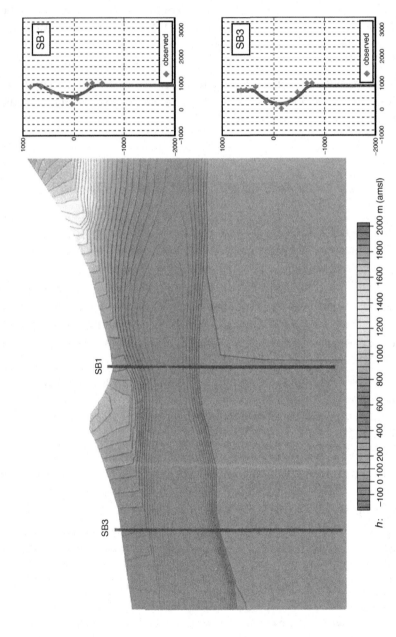

Fig. 5.75 Simulated hydraulic head distribution for today, i.e. $t = 10\,000$ years after the end of the last glaciation, showing match between computed and measured values in test wells SB1 and SB3 (Vinard, 1998, Fig. 9.4-2d, p. 135).

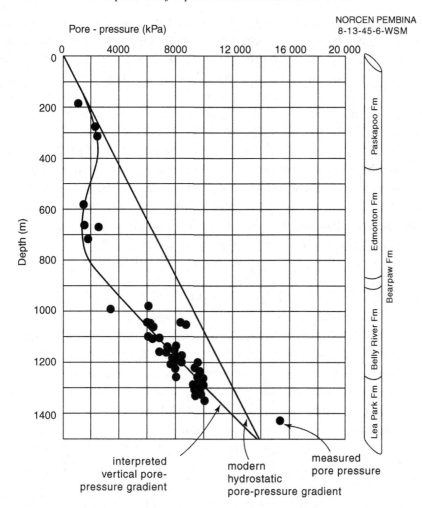

Fig. 5.76 Pressure-depth profile interpreted to indicate elastic dilation of the rock framework due to erosional unloading, west central Alberta, Canada (after Parks and Tóth, 1995, Fig. 4, p. 285).

application of hydrogeological principles to, and case studies of, exploration for petroleum, uranium and sulfide ores.

5.5.1 The hydraulic theory of petroleum migration and its application to exploration

The basic thesis of the hydraulic theory of petroleum migration (reviewed and illustrated in Section 4.4.2.6, Fig. 4.20) is that 'in geologically mature basins, gravity-induced cross-formational flow is the principal agent in the transport and accumulation of hydrocarbons. The mechanism becomes operative after com-paction of sediments and the concomitant primary migration cease, and subaerial topographic relief develops. Hydrocarbons from source or carrier beds are then moved along well-defined migration paths toward discharge foci of converging flow systems, and may accumulate en route in hydraulic or hydrodynamic traps. Accordingly, deposits are expected and observed to be associated preferentially with ascending limbs and stagnant zones of flow systems and hence to be charac-terized by relative potentiometric minima, downward increase in hydraulic heads possibly reaching artesian conditions, reduced or zero lateral hydraulic gradients and relatively high groundwater salinity' (Tóth, 1980, p. 121).

Figure 5.77 illustrates the salient points of the thesis. The large oil field in the Aquitaine region's Parentis Basin of France is located in a closed local minimum of the continental portion of the potentiometric surface in the Eocene age strata, indicating intensive flow into this area from all directions (Fig. 5.77a); the Cl^--content increases from less than 40 meq/l to over 700 meq/l in an areal configuration consistent with the potentiometric surface (Fig. 5.77b); and the groundwater flow pattern is cross-formational with local systems superimposed on the main regional system discharging into the Parentis Basin (Fig. 5.77c). Several other maps, not presented here, complete the picture by showing that the potentiometric surface in the Eocene reflects the area's regional topography (e.g., with the Pyrenees moun-tains to the south east), and that groundwater flow is ascending and artesian all along the coast, including the oil field (Tóth, 1980).

The concept of 'hydraulic trap', i.e. the common area of discharge between two or more concentrically converging gravity-driven groundwater flow-systems, was conceived originally for continental conditions (Fig. 4.20; Tóth, 1980, Fig. 44, p. 163). Tóth extended the concept later to include the region of 'convergence of con-tinental gravity flow and marine compaction flow of formation waters' (Fig. 5.78; Tóth, 1988, Fig. 11, p. 492).

In an extensive numerical analysis of the evolution of basinal flow conditions in the Gulf of Mexico, Harrison and Summa (1991) produced a sequence of quantita-tive illustration of opposing flow directions of, and intensive interplay between, continental meteoric and oceanic compaction waters (Fig. 5.79; Harrison and Summa, 1991, Fig. 12, p. 134).

To quote Harrison and Summa (1991, p. 109): 'We have defined three stages in the paleohydrologic evolution of the Gulf basin. In the first stage from Jurassic to early Tertiary times, the basin was characterized by circulating waters of meteoric origin, driven into the basin by topographic elevation episodically enhanced by eustatic falls in sea level. Abnormal pressures were restricted to the very deepest parts of the basin… In the second stage, during much of the Tertiary, geopressures developed beneath the major sedimentary depocenters related to the positions of the ancestral Mississippi and Rio Grande river systems. The circulation of meteoric

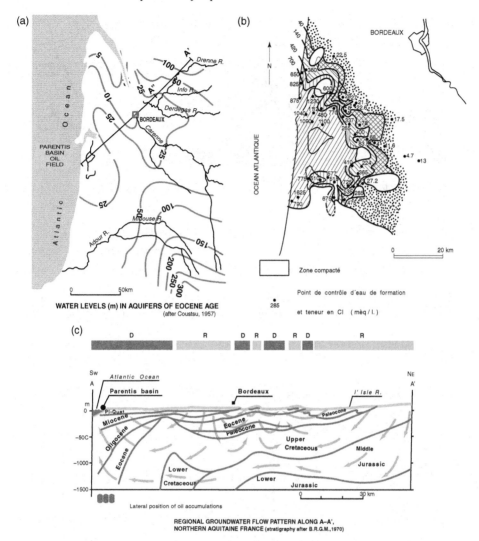

Fig. 5.77 Petroleum hydrogeologic conditions in the Parentis Basin, Aquitaine region, France: (a) potentiometric surface and oil fields in the Eocene-age aquifers; (b) formation water salinity in the Eocene-age aquifers; (c) groundwater flow pattern along cross-section A–A′ (after Tóth, 1980, Figs. a: 14, p. 137, c: 12c, p. 136, b: source unknown).

water became restricted by the opposing compactional hydraulic heads. The third stage in the development of Gulf basin hydrology began in late Miocene times and continues to the present day....'

Wells (1988) attributed the discovery(!) and emplacement of a major oil and gas accumulation offshore Qatar, Arabian Gulf, to hydrodynamic trapping between basinward flow of continental waters and landward movement of marine waters. The mixed waters were thought to be discharging together to the sea bottom through thinned out portions of shaley aquitards (Fig. 5.80; Wells, 1988, Figs. 1, 3, 8, 9,

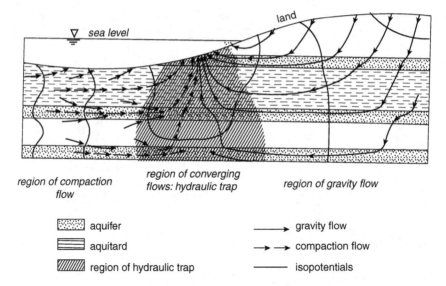

region of compaction flow

region of converging flows: hydraulic trap

region of gravity flow

aquifer

aquitard

region of hydraulic trap

gravity flow

compaction flow

isopotentials

Fig. 5.78 Hydraulic trap created by convergence of continental gravity flow and marine compaction flow of formation waters (Tóth, 1988, Fig. 11, p. 492).

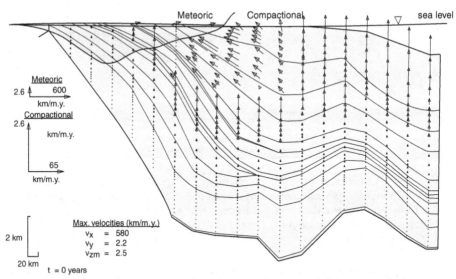

Computed regime of meteoric and compactional waters, present-day times, Gulf of Mexico (after Harrison and Summa, 1991)

Fig. 5.79 Flow vectors and fluid velocities due to compaction and topographic drive calculated for present-day along a N–S section through the northern Gulf of Mexico (Harrison and Summa, 1991, Fig. 12, p. 134).

pp. 358 and 360). The case is summarized in Figure 5.80, showing: the location of the general area and relevant site details (Fig. 5.80a); Wells' conceptual model of the local subsurface flow conditions (Fig. 5.80b); the potentiometric configuration in the reservoir beds and the petroleum accumulations in the closed regional potentiometric minimum (Fig. 5.80c); and the formation-water salinities increasing concentrically towards, and reaching a maximum in, the area of potentiometric minima and petroleum accumulations (Fig. 5.80d).

The Karamai oil fields in the Dzungarian Artesian Basin of NW China illustrate the possible close relations between gravitational groundwater flow systems, on the one hand, and the migration and accumulation of hydrocarbons together with various discharge-associated natural phenomena such as artesian wells and saline soils, and the 'geochemical chimney' (to be discussed below), on the other.

According to Tuan Yung-hou and Chao Hsuch-tun (1968), prior to the uplift of the Tien Shan mountains in pre-Quaternary times, the focus of groundwater discharge and petroleum accumulation occurred in the Tu-shan-tzu area in the south (Fig. 5.81a). The subsequent rise of the Tien Shan converted the Tu-shan-tzu area into a regional recharge region and induced strong northerly flow of groundwater. Most of the previously accumulated hydrocarbons were thus moved northward into the newly generated extensive discharge area in the deepest part of the Dzungarian Basin. The basin is characterized now by flowing artesian wells, saline soils, salt marshes and seeps, as well as springs of petroleum associated with the remigrated oil fields (Figs. 5.81b,c).

5.6 Potential role of flow-system analysis in surface geochemical prospecting

'Geochemical prospecting for petroleum is the search for chemically identifiable surface or near-surface occurrences of hydrocarbons as clues to the location of oil or gas accumulations. It extends through a range from observation of clearly visible oil and gas seepages at one extreme to the identification of minute traces of hydrocarbons determinable only by highly sophisticated analytical methods at the other' (Figs. 5.81c, 5.82; Hedberg, 1996, p. iii).

On the way to the surface from their source rock or reservoir, ascending hydrocarbon particles chemically reduce the traversed rock volume, the 'geochemical chimney'. The process may give rise to a wide variety of chemical, mineralogical, physical, pedological, botanical, microbiological anomalies, called collectively 'signatures' (Fig. 5.83).

The two mechanisms thought by most to be responsible for the migration of the oil and gas particles in the Geochemical Chimney are buoyancy and diffusion. Advection by moving groundwater has not been seriously considered. Consequently,

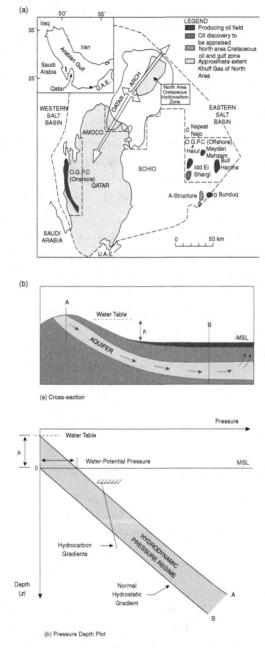

Fig. 5.80 Hydraulic trapping in the Cretaceous Nahr Umr sand, Offshore Qatar: (a) site map; (b) model of local hydrodynamics; (c) potentiometric map of the Shuaiba aquifer (one of the reservoir rocks); (d) formation-water salinities in the Nahr Umr and Shuaiba aquifers (Wells, 1988, Figs. 1, 3, 8, 9, respectively, p. 358 and 360).

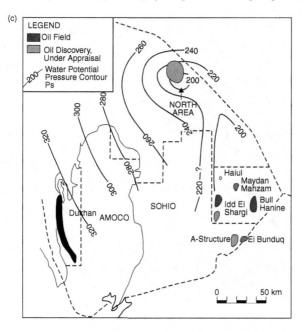

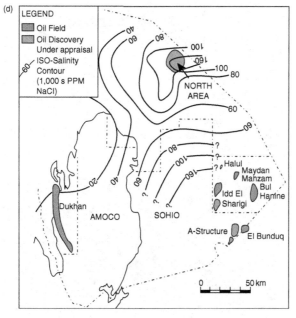

Fig. 5.80 (cont.)

(a)
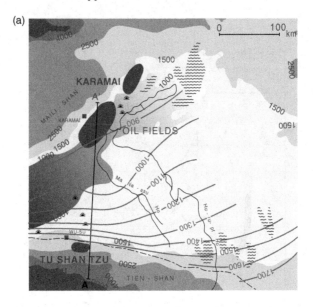

(b)
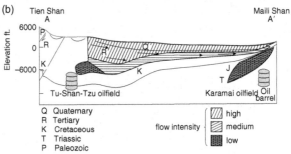

(c)

Fig. 5.81 Salient petroleum-hydrogeological features of the Karamai oil fields: (a) potentiometric surface with remigrated oil field in the new area of discharge and flowing wells; (b) hydraulic cross-section showing ascending water flow in the oil fields: (c) oil seep atop the oil fields [(a) and (b): from Tóth, 1980, after Tuan Yung-hou and Chao Hsuch-tun, 1968; (c) unknown source].

Fig. 5.82 J. Tóth (left) with a student at oil spring in Mae Soom Oil field, Fang Basin, north west Thailand.

the possible effects of groundwater flow on the location of geochemical signatures relative to the source of their petroleum components is commonly not taken into account, although their genetic and spatial relation is frequently observed and widely recognized (Fig. 5.84).

In an effort to dislodge the geochemical prospecting community from its state of conceptual inertia, the present author wrote (Tóth, 1996, p. 282):

'Near-surface exploration for petroleum was born more than 60 years ago. ...notwithstanding the diversity and advanced state of detection and analytical techniques, [it] still seems to yield inconsistent results not noticeably better than in its early days...

In my view, the main reason for this unsatisfactory and unnecessary state of affairs is that advective transport of hydrocarbons between reservoired sources and the near-surface anomalies is ignored in interpretation. Near-surface explorationists tend to accept the phrase 'vertical migration' literally [...] the effect of groundwater flow, which can shift, modify, or obliterate anomalies, is not considered or even recognized by most explorationists despite the sporadically but clearly stated warning found in the literature... Hydrogeology is well equipped, both theoretically and technically, to evaluate site-specific situations' (Tóth, 1996, p. 282–283). I consider this message still to be valid. Perhaps the hydrogeological community should take an active role in convincing the surface geochemical explorationists.

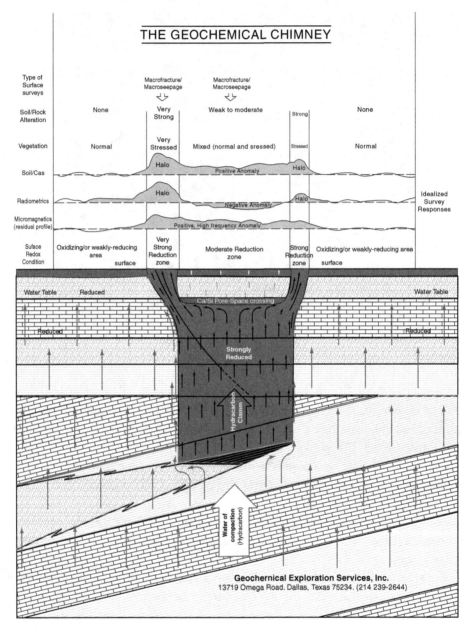

Fig. 5.83 Commercial illustration of the 'geochemical chimney' from the 1980s.

An example of the potential economic value of flow-system analyses in geochemical prospecting, albeit on the money-saving side of the ledger, is the Holysh–Tóth (1996) study concluding that traditional geochemical prospecting is likely to be ineffective in recharge areas of gravity-driven groundwater flow.

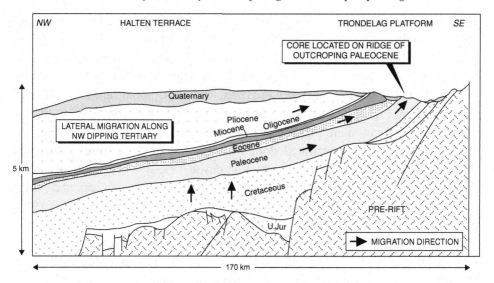

Fig. 5.84 Geologic cross-section across the Halten Terrace and Trondelag plat-form, Central North Sea: 'Vertical leakage from Jurassic reservoirs and source rocks is displaced laterally [by] up-dipping sandier beds in Paleocene deposits and is found at the surface displaced as far as 50 km from the fields' (Thrasher *et al.*, 1996, Fig. 14, p. 235; reprinted by permission of the AAPG whose permission is required for further use).

5.6.1 Exploration for uranium deposits by groundwater flow-system analysis

Hagmaier (1971) may have been the first to suggest that analysis of groundwater flow-systems might aid uranium exploration: 'Hostetler and Garrels' (1962) investi-gation into the geochemistry of uranium in water demonstrated that it is chemically possible for groundwater to transport, deposit, and concentrate uranium. It should be noted, however, that most of the literature concerning the role of uranium in groundwater in uranium deposition does not fully utilize recent developments in groundwater geology' (Hagmaier, 1971, p. 19).

Hagmaier (1971) proceeded to propose a working hypothesis as a possible strategic basis of exploration for uranium by combining those 'recent develop-ments', concerning the geometrical aspects of groundwater flow patterns (Tóth, 1962a, 1963) and the associated distribution of chemical water types (e.g., Meyboom, 1966), with the (then) known locations, geological history, mineral-ogy, and geochemistry of uranium deposits of the Powder River Basin, Wyoming, USA. According to his working hypothesis, bicarbonate-rich groundwater of the recharge areas is capable of leaching uranium as uranyl carbonate complexes as well as transporting them into depths where chemical reduction and changes to a sulfate facies of the chemical water-type precipitate uranium. The uranium carried in the carbonate complexes would thus likely be associated with the bicarbonate chemical facies found in recharge areas (Fig. 5.85; Hagmaier, 1971, Fig. 4, p. 22).

In a study based on more than a decade of relevant prior experience, and on the geological history and advanced modelling of groundwater flow in the San Juan basin of the Four Corners area (Fig. 5.86; Utah, Arizona, Colorado and New Mexico; Sanford, 1994, Fig. 1, p. 341), Sanford concluded: 'The most important results for uranium ore formation are that regional ground water discharged throughout the basin, discharge was concentrated along the shore line or playa margin, flow was dominantly gravity driven, and compaction dewatering was negligible. A strong association is found between the tabular sandstone uranium deposits and major inferred zones of mixed local and regional ground-water discharge' (Sanford, 1994, p. 341).

Regarding the likely location of uranium deposits within the groundwater flow pattern, Hagmaier (1971) and Sanford (1994) have come to opposite conclusions. Nevertheless, a fundamental common denominator, shared by several other students of the question (see Section 4.4.2.6), is present in both opinions, namely, that the location of roll-front and tabular type uranium deposits are genetically bound to identifiable segments of gravity-driven groundwater flow systems. Invoking also the power of numerical modelling techniques, Sanford (1994, p. 357) concludes with a persuasive summary: 'In light of this progress, modern methods of paleohydrologic analysis should be considered as a standard tool in the analysis of a wide variety of ore deposit types as well as in the exploration for new deposits.' (See the conclusion of Baskov (1987) in Section 5.5.4.)

5.6.2 *Gravity-driven flow systems and strata-bound ore deposits*

Knowledge of quantitatively defined physical and chemical conditions required for the formation of ore deposits helps to improve the efficiency of exploration by limiting the expectable areas of their occurrence. Garven and Freeze (1984) were among the first researchers to apply rigorously formulated sensitivity analyses of gravity-driven groundwater flow to the problem of the genesis of strata bound sulfide ore deposits, triggering an explosion of interest in the topic. They illustrated their modelling approach and the essential character of the results by figures exemplified in Figure 5.87, with the warning: 'it should not be taken as the final result.' (Garven and Freeze, 1984 Pt. 1. Fig. 8, p. 1108). Figure 5.87(a) shows the distribution of gravity-induced hydraulic heads calculated with water density, viscosity, salinity and temperature taken into account, and the associated flow velocities, along a cross-section of a hypothetical basin. The change in location and intensity of a mass-concentration pulse is shown at three different points in time in Figure

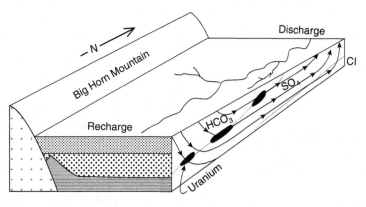

Fig. 5.85 Conceptual model of uranium deposition in gravity-driven groundwater flow system in the Powder River Basin, Wyoming, USA (Hagmeier, 1971, Fig. 4, p. 22).

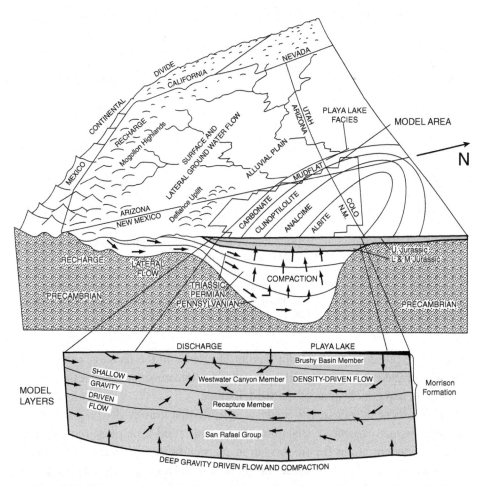

Fig. 5.86 Conceptual model of groundwater flow in the Late Jurassic to Early Cretaceous San Juan basin, New Mexico, USA (Sanford, 1994, Fig. 1, p. 342).

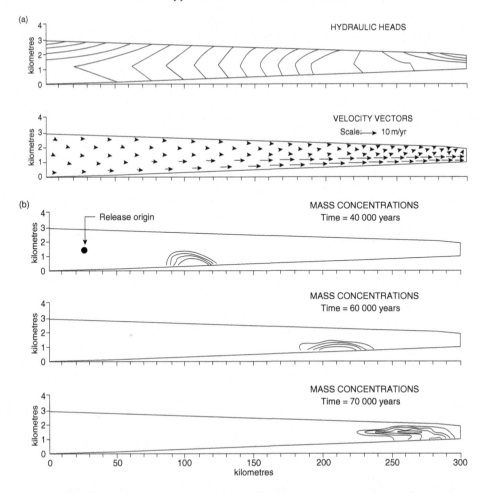

Fig. 5.87 Illustrative sample computation of gravity-driven groundwater flow and mass transport from a point source of metal-bearing non-reactive pulse in a generic drainage basin: (a) distributions of hydraulic head and flow velocity; (b) changes in location and intensity of the migrating pulse with respect to time, with the contours indicating 3, 6, 9 and 12 per cent of solute concentration at the source (excerpt from Garven and Freeze, 1984, Pt. 1, Fig. 8, p 1113–1114).

5.87(b). The pulse originates from a point of release in the recharge area and the contours indicate the 3, 6, 9 and 12 per cent of solute concentration at the source.

The principal conclusions of the work include:

'The results of the sensitivity analysis indicate that gravity-driven groundwater flow systems are capable of sustaining favourable fluid-flow rates, temperatures, and metal concentrations for ore formation at groundwater discharge areas near the edge of the basin... Long-distance transport of metal and sulfide in the same fluid can probably be ruled out largely on hydrogeochemical grounds but also because

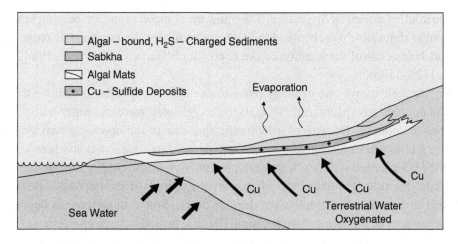

Sabkha – diagenetic model for copper mineralization to explain the origin
of the Zambian Copperbelt and other intracratonic (diagenetic) Cu deposits.

Fig. 5.88 Model of copper-sulfide deposition resulting from the mixing of copper carrying oxygenated (fresh) water, saline sea water, algae-produced H_2S (modified from Morganti, 1981).

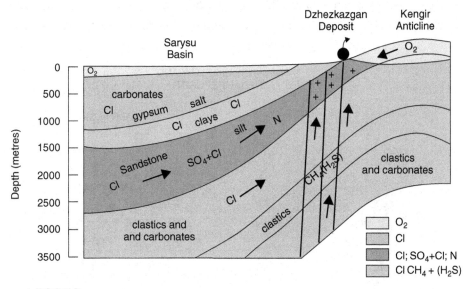

Fig. 5.89 Copper and lead–copper deposit generated in the common paleo-discharge zone of oxygenated meteoric waters and ascending chloride- and $CH_4 - H_2S$-charged chloride brines, Dzhezkazgan, former Soviet Union (modified from Baskov, 1987, Fig. 48, p. 184).

of the diluting effects of dispersion. The transport of metal in sulfate-bearing brines is a more defensible geochemical model, in which case the presence of reducing agents helps control the location of ore deposition' (Garven and Freeze, 1984, Pt. 2, p. 1125–1126).

Figure 5.88 shows one of many real-life examples supporting the above theoretical conclusions (Morganti, 1981). Here, oxygenated meteoric water transports copper in solution. By way of a hydraulic trap due to the opposing heavier sea water, it is forced to discharge upwards into the sabkha's algal mat and is reduced by the H_2S-charged sediments, causing precipitation of the copper sulfide.

Under the title 'Fundamentals of paleohydrogeology of ore deposits', Baskov (1987) devotes an entire book to the detailed description of numerous ore deposits in the territory of the former Soviet Union. Because he finds a large proportion of the ore deposits associated with paleo-discharge centers of groundwater, Baskov puts special emphasis on their recognition and mapping: 'Locating the discharge regions (centres) of ancient ground waters is an important part of paleohydrogeological studies. Among direct indications of ancient ground-water discharge centres, various near-surface vein formations (ore formations inclusive) may be cited, as well as zones of metasomatic rocks, travertine deposition and other traces of groundwater activity related to its discharge onto the ground surface' (Baskov, 1987, p. 104). The example of the Dzhezkazgan copper and lead–copper deposit shows chloride waters rich in methane (CH_4) and hydrogen sulfide (H_2S) ascending along a highly permeable subvertical fault zone into the area where fresh meteoric water

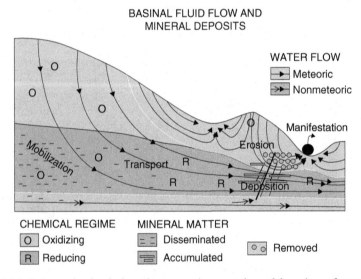

Fig. 5.90 Schematized relations between the genesis and location of stratiform ore deposits and gravity-driven groundwater flow systems.

and sulfate–chloride brine converge, with metals precipitating in the reductive zone of the mixing liquids (Fig. 5.89). Baskov draws the conclusion that the application of 'paleohydrogeology' to prospecting for ore deposits consists of prospecting for discharge regions of 'artesian basins' of paleo-groundwater flow systems.

Figure 5.90 presents a modern hydrogeologist's conceptualization of the genetic relations between the location of stratiform ore deposits on the one hand, and gravity-driven groundwater flow systems, on the other.

6

Epilogue: gravitational systems of groundwater flow and the science of hydrogeology

Hydrogeology can be defined today as the science and practical utilization of those processes and phenomena that result from the interaction between groundwater and the rock framework. However, with its broad scope and multifarious applicability modern hydrogeology is fundamentally different from what it was before the 1960s. In those days, it was a single-issue subject of providing water supplies from subsurface sources.

The multitude and diversity of the various, and in some cases seemingly unrelated, processes, phenomena and issues presented in the preceding pages may obscure the fundamental and abrupt change in the scope of hydrogeology caused by the concept of gravity-driven groundwater flow systems. The intent of this chapter is, therefore, to put the previous five chapters into a concise historical context. To this end, the chronological evolution of some groundwater related conjectures, theories, knowledge and technical approaches will be reviewed briefly from the perspectives of the natural sciences and engineering sciences.

Hydrogeology matured suddenly into a fully fledged member of the earth sciences during the remarkably short period of approximately 30 years between the late 1950s and early 1980s. Prior to 1960, the discipline was the realm chiefly of the geologist, a natural scientist with little or no interest and background in the quantitative statements of the laws and factors controlling the flow of groundwater, let alone the differential equations describing them. On the other hand, in the work to develop groundwater supplies, the hydraulics engineer was unable to deal with the grey area between 'permeable' and 'impermeable' geologic formations and wrote the well-yield equations for aquifers with open water tables, or for strata sandwiched between ideally impermeable aquicludes. Arguably, the processes of rapid conceptual evolution and broadening scope were triggered by the virtually simultaneous recognition of the rock framework's regional hydraulic continuity in the two principal sub-disciplines interested in groundwater, namely, hydrogeology, and aquifer-and-well-hydraulics.

244

The two-pronged approach to groundwater-related questions started in times immemorial. Evolution continued well-nigh independently side by side along parallel paths until the 1950s when, driven by the scientist's curiosity, on the one hand, and the engineer's practicality, on the other, the branches began to converge. However, both branches can also be divided into two subsequent segments of time for this period. First, the earlier stages, when explanations of water-related natural processes and phenomena were based on speculation and techniques of capturing groundwater were developed empirically. These stages were followed by the times when theories and hypotheses started to be expressed quantitatively and with mathematical rigour (Fig. 6.1).

Along the 'natural sciences branch', the segment of 'speculative explanations' stretched into the seventeenth century. The first recorded questions, and answers, during this period pertained to the origin of springs, and the hydrologic cycle (Meinzer, 1942). Great thinkers, starting with Homer in the eighth century BC, and including Thales (ca. 640–546 BC), Plato (427–347 BC), Aristotle (384–322 BC), and even Johann Kepler (1571–1630) and René Descartes (1596–1650) speculated that springs issued from water squeezed out of the sea, or from condensation collected in caves. Meteoric water was explicitly considered to be insufficient to feed the rivers. In another camp, though, Marcus Vitruvius Pollo (86–20 BC) thought that springs derived from infiltrated rain. This view was supported later

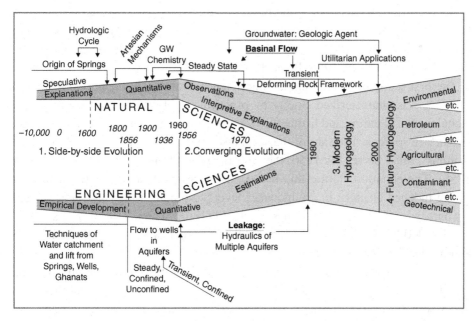

Fig. 6.1 Overview of the historical evolution of the science of hydrogeology.

by Leonardo da Vinci (1452–1519) and, in 'Discours admirable de la nature des eaux et fontaines tant naturelles qu'artificielles' (Wonderful discourse on the nature of waters and fountains both natural and artificial; 1580, Paris) by Bernard Palissy (1510–1589).

Quantitative hydrological observations began in the seventeenth century with Pierre Perrault (1608–80). He measured rainfall for three years in the Seine river basin and showed that it was six times the quantity discharged by the river. Edmond Mariotté (1620–84) verified the results of Perrault, and Edmund Halley (1656–1742) demonstrated that evaporation could account for runoff into the Mediterranean. Jean-Claude La Métherie (1743–1817) started to measure rock permeabilities and also divided infiltration into surface runoff and deep reservoirs (1791), thus conceptualizing a rudimentary water balance.

Although the first flowing well in Artois was drilled in 1126, the first recorded explanations of the artesian phenomenon appear only in the seventeenth century from Giovanni Domenico Cassini (1625–1712) and Antonio Vallisnieri (1661–1730), both of whom correctly identified high water pressure in confined aquifers as the cause. Further attempts to refine the concept (Chamberlain, 1885) may have contributed to strengthening the notion of absolute impermeability. Nevertheless, the attention paid to regionally extensive water-bearing and confining strata has probably, in itself, led to the notion of a groundwater basin (e.g. Slichter, 1902), with the most fundamental and influential work in this direction being *The Theory of Ground-Water Motion* by M. King Hubbert (1940).

Studies in the chemistry of groundwater, for other than solely utilitarian purposes, began in the late nineteenth–early twentieth century focusing on classification (Palmer, 1911; Schoeller, 1935, 1955, 1962) and on factors of compositional evolution (Chebotarev, 1955; Back, 1960).

The first segment of the Engineering Science Branch, i.e. the period from $-\infty$ to 1856, was characterized chiefly by practical development of implements and techniques of catching and lifting the water from springs, wells, ghanats, and other subsurface sources. The start of the second segment, that of the 'Quantitative Estimations', must be set for 1856, the year of publication of Darcy's Law (Darcy, 1856). The equation stating this law triggered an explosion of interest in measuring induced changes in groundwater levels to predict well yields and it led to the numerous well-known formulas of Dupuit (1863), Thiem (1906), Theis (1935), Jacob (1940), Wenzel (1942), and so on, all related to flow to wells in the water-table and/or ideally-confined aquifers.

Whether by fortuitous coincidence or subconscious cross-fertilization during the mid 1950s and through the 1960s, the notion of absolute impermeability came under serious scrutiny by both engineers in their evaluations of aquifer and well yields, and by geologists in their studies of groundwater flow in drainage basins.

Hantush and Jacob (1954), Hantush (1956), Witherspoon and Neuman (1967), Margat (1969), and others introduced and developed the concept of leaky aquifers and expanded it into the concept of multiple-aquifer systems of basinal scale (Neuman and Witherspoon, 1971, 1972). In the natural sciences branch, Tóth's (1962a, 1963) homogeneous 'unit basin' and composite basins of sinusoidal water table were 'heterogenized' by Freeze and Witherspoon who numerically modelled the effects on basinal flow patterns of aquifers and aquitards of various attitudes and extent (Freeze and Witherspoon, 1967). The theoretical results were supported by published field studies (some early examples: Geraghty, 1960; Pluhowsky and Kantrowitz, 1964; Kolesov, 1965; and others) and/or purposefully field-tested (Tóth, 1966b; Mifflin, 1968). The final conclusion of both the engineering and natural science sides was: *The rock framework is hydraulically continuous* (Tóth, 1995).

As a direct consequence of that conclusion, it was soon recognized that hydraulic continuity results in the development of flow systems on widely different scales of space and time (Zijl, 1999), and that the systems at each scale may generate natural processes and phenomena of characteristic type and form. *The unifying notion that flowing groundwater is a geologic agent was born* (Tóth, 1999).

The convergence of the natural and engineering science branches may be considered to have been completed by approximately 1980, ushering in the age of the mature Modern Hydrogeology. This new member of the earth sciences is both a basic discipline and a specialty. Familiarity with its basic tenets has become indispensable for thorough and effective work in most all other geological and geotechnical activities. At the same time, its specialized aspects require a unique educational and professional background and full-time commitment of the hydrogeologist.

Modern hydrogeology can be characterized by two principal paradigms:

(i) gravity-driven groundwater flow occurs on broad spectra of spatial and temporal scales;
(ii) flowing groundwater is a ubiquitous geologic agent responsible for a broad spectrum of natural processes and phenomena at and to great depths below the surface of the Earth.

Whither hydrogeology? Having become a mature science, built upon the mathematical rigour of the engineer and the licentious imagination of the earth scientist, as well as on all the associated skills, methods and techniques that those disciplines employ, major developments or breakthroughs in hydrogeological concepts, theories or techniques are unlikely to happen in the foreseeable future. Instead, increasing specialization can be expected with growing numbers of 'qualified' or hyphenated sub-disciplines appearing, such as: environmental-, contaminant-, agricultural-, eco-, petroleum-, and so on, hydrogeologies (Fig. 6.1).

Glossary

Absolute anomaly (of pressure, fluid potential or hydraulic head): the difference between the perturbed and the unperturbed fields at a given point in the flow domain caused by a rock body, where the permeability is different from its homogeneous surrounding.

Absolute pressure: pressure referenced to vacuum, i.e. to absolute zero pressure.

Adjustment time: the length of time a change in energy (pore pressure, hydraulic head) at one point in the flow domain reaches a specified percentage of the total original change initiated at another point. (see **characteristic time**).

Anisotropic: property of rock having anisotropy.

Anisotropy: the property of having different magnitudes of hydraulic conductivity (permeability) in different directions.

Anomaly (of **fluid potential, pressure,** or **hydraulic head**): The difference between the actual perturbed fields and the unperturbed original fields of fluid potential, pressure, or hydraulic head at a given point in the flow domain (caused not necessarily by permeability differences in the rock framework).

Aquifer: a relatively highly permeable rock unit, or group of rock units, capable of supplying water at required rates.

Aquifer and well hydraulics or **aquifer engineering:** the study and practice of water withdrawal from, or injection into, geologic formations by means of boreholes, galleries, or other engineering facilities for utilitarian purposes.

Aquifer engineering see **aquifer and well hydraulics.**

Aquitard: a relatively poorly permeable rock unit, or group of rock units, incapable of supplying water at required rates.

Artesian: condition in which the potentiometric surface is above the top of the aquifer to which it belongs. Used also, incorrectly, to mean **'flowing'** conditions, i.e. in which the potentiometric surface is above the land surface.

Atmosphere: the gaseous mass, including water vapour, enveloping the Earth.

Atmotrophic: condition of a site poor in plant nutrients (low trophic levels owing to short subsurface flow paths of infiltrated atmospheric water; commonly associated with groundwater recharge areas).

Base flow: the annual minimum discharge of a stream.

Basin(al) hydraulics: the spatial distribution patterns and intensity of groundwater flow at the scale of drainage basins and/or their study for utilitarian and scientific purposes.

Bog: an area of organic soil, mostly peat, commonly occurring in groundwater recharge areas and characterized by acidic (pH<7) and oxidizing chemical, and atmotrophic (poor) plant nutrient conditions.

Boundary condition see **boundary value**.

Boundary value or **boundary condition:** magnitude and/or direction of fluid-dynamic parameters specified along the boundaries of a flow domain.

Boundary value problem: mathematical task to determine the properties of a groundwater flow-field in a flow domain (distributions of fluid potential, flux, pressure, temperature, hydraulic gradients, and so on) from boundary conditions.

Breakthrough time or **onset time:** point in time at which a change in pore pressure (hydraulic head) induced at a point of origin in the flow domain arrives, i.e. becomes observable, at a point of observation (applied also to arrival of heat or matter transported in groundwater, e.g. chemical constituents, contaminants).

Brine: saline formation water with total dissolved salt content exceeding that of sea-water, i.e. approximately 35 000 mg/l.

Characteristic time: duration of transient condition of the water table (or flow system, controlled by it) while changing from an initial position to the final stabilized state induced by a constant specified change in recharge (100 per cent of **adjustment time**).

Composite drainage basin: drainage basin with a water table of sufficient local relief to induce multiple flow systems.

Composite flow pattern: flow pattern comprising several flow systems (common consequence of water tables of high relief, as in **composite drainage basins**).

Conductive heat transport: propagation of heat by diffusion.

Confined aquifer: aquifer bounded by rock units of contrastingly lower permeability; if saturated, pore pressure at the top of a confined aquifer is equal or greater than hydrostatic.

Conservative flow field: flow field without internal sources and sinks: the rate of mass recharge equals the rate of mass discharge through its boundaries.

Continuity equation see **equation of continuity**.

Convective heat transport: propagation of heat by flow of a fluid, e.g. water.

Coulomb failure potential: a dimensionless measure of the probability of shear failure in a slope. It is defined as the ratio of the maximum shear stress and the mean normal stress and is considered positive in tension, it is independent of the material strength and has a range from zero to one.

Darcy's law: the basic relation between the magnitude and orientation of the fluid-driving force and the rate and direction of the resulting flow through permeable media.

Darcy velocity see **Specific volume discharge**.

Datum plane: horizontal surface, $z = 0$, arbitrarily chosen as reference for measurement of elevation; sea level is commonly used in hydrogeology.

Diffusion (of energy) or **Pressure diffusion** or **Pore-pressure diffusion:** process or time-rate of dissipation of the water's mechanical energy. It is controlled by the fluid-potential gradient and the **diffusivity coefficient,** and is manifest by temporal changes in pore pressure or hydraulic head in the flow domain (definition is applicable also to other forms of energy, e.g. thermal).

Diffusion equation: differential equation relating the spatial and temporal distributions of hydraulic heads and flow inside a flow domain to the rock framework's hydraulic conductivity and specific storage and to the initial values of dynamic parameters (e.g., hydraulic head) on the domain's boundaries.

Diffusivity see **hydraulic diffusivity**.

Diffusivity coefficient (hydraulic): measure of hydraulic diffusivity expressed as the ratio of hydraulic conductivity and specific storage.

Dimension: parameter used singly or in combination with others to characterize physical properties and conditions of matter, such as: length $[L]$, area $[L^2]$, volume $[L^3]$, mass $[M]$, time $[T]$, velocity $[L/T]$, acceleration $[L/T^2]$, density $[M/L^3]$, pressure gradient and specific weight $[M/T^2L^2]$ (see also **unit of measurement**).

Discharge see **groundwater discharge**.

Discharge area see **groundwater discharge area**.

Discharge feature: observable emergence of groundwater to the land surface or the consequence of it resulting from the physical, chemical, or dynamic interaction of the water with its environment.

Disequilibrium: condition of a single value or a field of parameters (e.g., pressure, velocity, temperature, chemical constituents) in which the values are not adjusted to their domain's boundary conditions.

Drainage basin: an area of the land surface that gathers water of precipitation and contributes it to a particular stream channel or system of channels.

Drainage divide see **water divide**.

Drill stem test (DST): technique of pore-pressure measurement and formation-fluid sampling in boreholes by means of a set of valves and ports replacing the drill bit on the end of the hollow rod of a drilling machine.

Dynamic head increment: the difference in hydraulic heads between the actual (i.e. real or dynamic) and nominal (assumed static-state) values representing the conditions of the water's mechanical energy in the flowing and, respectively, static states in one single subsurface point.

Dynamic pressure gradient: difference in pore pressure over a unit vertical length in a body of moving groundwater.

Dynamic pressure increment: the difference between the actual (real or dynamic, $p_{real} = p_{dyn}$) and the nominal (assumed static-state, $p_{nom} = p_{st}$) pressure, representing the pore-pressure values in the flowing and, respectively, static state in one single subsurface point.

Eco-hydrological: of, or, pertaining to **eco-hydrology**.

Eco-hydrology: hydrology applied to ecological problems; specifically, the study of the relationship between the type and quality of plant associations and the chemical conditions and water availability at their site.

Ecology: the study of the relationships between organisms and their environment including communities, patterns of life, population changes.

Ecosystem: a unit in ecology comprising the environment with its living elements, plus the nonliving factors that exist in and affect it.

Effective stress: pressure between solid component elements (grains, crystals, fracture surfaces, fragments) of soils and rocks.

Effectively impermeable (effectively impervious) base see **penetration depth**.

Elevation head: height of a point of observation relative to the datum plane.

Epilimnic or **epilimnetic** or **epilimnial:** condition (physical, chemical, thermal, biological) in the **epilimnion**.

Epilimnion: the uppermost layer of water in a lake characterized by relatively high oxygen content and uniform and high temperature as compared with deeper zones in the lake.

Equation of continuity or **continuity equation:** mathematical statement of the principle of mass conservation, i.e. the rate of mass flow into a control volume equals the rate of mass flow out of it.

Equipotential line or **equipotential:** line along which the value of fluid potential (hydraulic head) is constant.

Equipotential see **equipotential line**.

Eutrophication: anomalously rich growth of aquatic vegetation due to above normal influx of nutrients.

Exfiltration zone see **groundwater discharge area**.

Fen: area of organic soil, mostly peat, commonly occurring in groundwater discharge areas and characterized by alkaline (pH >7) reducing chemical, and LITHOTROPHIC (rich) plant-nutrient conditions.

Flow domain: area or volume within the boundaries of which a flow problem is considered.

Flow field see **flow pattern**

Flow-generated or/and **flow-sensitive natural processes and phenomena:** natural conditions the creation or modification of which would not occur without the effect of gravity-driven groundwater flow (do not include similar conditions not resulting from groundwater flow). Some examples: potentiometric anomalies; patterns of chemical composition of groundwater; soil salinization; soil mechanical weaknesses, liquefaction ('soap holes', 'quick sand', instability of slopes); anomalies in type and quality of vegetation; uranium and sulfide ore deposits; petroleum accumulations; geothermal anomalies; distribution and type of wetlands; and so on.

Flowing (as in: condition) see **artesian**.

Flowing well: well in which the water level rises naturally above the land surface.

Flow line: imaginary line representing a flow path.

Flow net: two-dimensional graphical portrayal of the flow field by the two intersecting sets of equipotential lines and flow lines. The two sets are orthogonal in isotropic media and generally non-orthogonal in anisotropic media.

Flow path: loci of points in space through which a volume element of fluid moves.

Flow-path volume: volume of rock encompassed by a plume of water transported by flowing groundwater from a source such as a repository of radioactive fuel-waste emanating from a repository, see **plume**.

Flow pattern or **flow field:** spatial distribution of fluid flow in a flow domain, commonly represented by flow lines, flow vectors and, possibly, equipotential lines.

Flow regime: part or the whole of a basinal flow-field of groundwater in which the flow of water is uniformly one-directional with respect to the water table, i.e. descending (recharge) or parallel (transfer) or ascending (discharge).

Flow-sensitive natural processes and phenomena see **flow-generated** or/and **flow-sensitive natural processes and phenomena**.

Flow strength see **specific volume discharge**.

Flow system: see **groundwater flow system**.

Flow vector: arrow symbol orientated in the direction of the movement of water, with its length being proportional to the flux, or one of its components.

Fluid-dynamic parameter: a quantitative expression of mechanical energy and/or a property of the flow field, such as: fluid potential, hydraulic head, pore pressure, hydraulic gradient, flux, and so on.

Fluid potential (for incompressible fluids), $\Phi = gh = gz + p/\rho$: the work performed on, or by, a unit mass of fluid, i.e. the amount of mechanical energy contained by a unit mass of fluid, where: Φ, fluid potential; h, hydraulic head; g, acceleration due to gravity; z, elevation above datum; p, gauge pressure (absolute pressure minus atmospheric pressure); ρ, density of water.

Fluid-potential anomaly see **anomaly**.

Flux see **specific volume discharge**.

Formation pressure see **pore pressure**.

Gauge pressure: Pressure referenced to atmospheric pressure, i.e. to one atm., or to approximately 100 kPa.

Geochemical chimney: a chemically reduced column of earth material marked by any or any combination of a variety of chemical, physical, mineralogical, microbial, and vegetal anomalies, called signatures, attributed to hydrocarbon particles migrating from a natural accumulation of petroleum towards areas of lower fluid potential as, for instance, the land surface or sea bottom.

Geochemical exploration: exploration for petroleum based on observations of conditions attributed to the **geochemical chimney**.

Geopressure: pore pressure due to the weight of the overlying saturated column of rock.

Geosphere: the solid portion of the Earth, lithosphere; see **rock framework**.

Ghanat: tunnel with lined walls dug in the central portions of arid valleys in Persia (Iran) collecting groundwater from adjacent mountains.

Gravity-driven groundwater flow: subsurface movement of water through the rock's openings induced by differences in elevation of the water table.

Groundwater discharge: process by which water is removed from the saturated zone, or the rate or amount of water thus removed.

Groundwater discharge area or **discharge area, exfiltration zone, outflow region:** a delimited expanse of land surface beneath which groundwater near below the water table moves upwards, i.e. towards, the water table.

Groundwater divide: elevated area of the potentiometric surface from which groundwater moves in diverging directions.

Groundwater flow: movement of water through the rock's openings below the land surface.

Groundwater flow pattern: set of flow lines depicting the spatial distribution of flow paths in a subsurface flow domain.

Groundwater flow-system or **flow system:** a family of flow lines connecting the whole or part of a recharge area with the whole or part of a discharge area.

Groundwater flow system – intermediate or **intermediate (groundwater) flow system:** a family of flow lines connecting a recharge area with a discharge area that is not contiguous with it within a specified drainage basin.

Groundwater flow system – local or **local (groundwater) flow system:** a family of flow lines connecting a recharge area with a contiguous discharge area within a specified drainage basin.

Groundwater flow system – regional or **regional (groundwater) flow system:** a family of flow lines connecting the principal recharge area (water divide) with the principal discharge area (main valley) of a specified drainage basin.

Groundwater recharge: process by which water is added to the saturated zone, or the amount of water thus added.

Groundwater recharge area or **inflow region** or **infiltration zone** or **recharge area:** a delimited expanse of land surface beneath which groundwater near below the water table moves downwards, i.e. away, from the water table.

Groundwater regime: the six main attributes taken together and characterizing groundwater conditions in a region: (1) water content of the rocks; (2) geometry of the flow systems; (3) specific volume discharge; (4) chemical composition of the water; (5) temperature; and (6) the variations of all these parameters with respect to time.

Groundwater: water contained in the rock's voids below the land surface.

Halophyte: plant preferring saline soil conditions.

Heather: a plant (*Calluna vulgaris*) of the family Ericaceae and/or an extensive area of rather level, open, and uncultivated land with poor soil, inferior drainage, and a surface rich in peat or peaty humus.

Heterochronous: adjective applied to a flow field comprising flow systems of different ages.

Heterogeneity: the quality of being heterogeneous.

Heterogeneous: rock framework or flow domain in which the hydraulic conductivity is variable in space.

High-level nuclear-fuel waste or **high-level radioactive waste:** the highly radioactive material resulting from the reprocessing of spent nuclear fuel, including liquid waste produced directly in reprocessing and any solid material derived from such liquid waste that contains fission products in sufficient concentrations [§ 63.2 Definitions, Nuclear Waste Policy Act, 1982; United States Nuclear Regulatory Commission (USNRC)].

Homogeneity: the quality of being homogeneous.

Homogeneous: rock framework or flow domain in which the hydraulic conductivity is constant in space.

Hydraulic barrier: part of the rock framework with a contrastingly lower permeability (of various possible origins: structural, sedimentological, diagenetic, etc.) than that of the adjacent rock units.

Hydraulic conductivity: property of the rock–water system measured by the volume of water flow through, and normal to, a unit surface of permeable material during a unit length of time under a unit difference in hydraulic head over a unit distance. It is dependent on the water's viscosity and density, thus on temperature.

Hydraulic conduit: part of the rock framework with a contrastingly higher permeability (of various possible origins: structural, sedimentological, diagenetic, etc.) than that of the adjacent rock units.

Hydraulic continuity: scale-dependent empirical property of the rock framework expressed as the ratio of an induced change in hydraulic head (pore pressure) at a point of observation to an inducing change in hydraulic head (pore pressure) at a point of origin. Its theoretical value ranges from zero to one and it depends on the distance between the points of origin and observation, the path and speed of pressure propagation (i.e. spatial distribution of hydraulic conductivity and diffusivity), and the timing and duration of observation. Thus it is a function of the scales of space and time at which a given problem is treated.

Hydraulic diffusivity: property of a saturated compressible rock framework controlling the rate of energy dissipation (pressure, hydraulic head) in response to a change in boundary conditions and expressed as the ratio of hydraulic conductivity and specific storage.

Hydraulic exploration (for reservoir quality rock pods): search for potentiometric anomalies caused by the flow of groundwater through rock bodies of relatively high permeability.

Hydraulic gradient: change in hydraulic head over a unit length of flow path, a vector, taken positive in the direction of increasing hydraulic head.

Hydraulic head or **potentiometric elevation:** the height of a vertical water column relative to an arbitrary datum plane commonly chosen at sea level. It is directly proportional to, thus a measure of, the fluid potential in a given point of the rock framework.

Hydraulic trap: closed concentric, or linear, regional minimum of the fluid potential field in the *XY* plane inducing horizontally converging flow of formation water. (Petroleum hydrocarbons, if present, migrate towards the centre of the closed potentiometric minima or troughs and, where water flow turns vertical, normally upward, they are retained by a combination of slightly permeable strata, reduced temperature and pressure, and increased water salinity.)

Hydraulically continuous: condition of the flow domain between two point of which the hydraulic connection is greater than zero.

Hydrodynamic (oil, gas) **trap:** local combination of the fields of permeability and fluid potential (i.e. flow) effecting or enhancing the separation of petroleum hydrocarbons from water, and their local retention and accumulation (common in anticlines, monoclines, pinch-outs, lenses).

Hydrogeologic environment: collective term for the three components of a geographic region, namely, topography, geology, and climate, that together control the area's groundwater conditions i.e. the 'Groundwater Regime'.

Hydrogeological reconnaissance map (of Alberta, Canada): Small-scale map (M =1:125 000 to 1:500 000) showing components of the **hydrogeological environment** and parameters of the **groundwater regime**. (Their intended use is to enable the evaluation of hydrogeologic conditions on regional scales to be used for specific purposes according to need, as well as to serve as a guide for effective and efficient continuation of hydrogeological mapping at larger scales http://www.ags.gov.ab.ca/publications/).

Hydrological (or **water-**) **system** (Engelen and Kloosterman, 1996, p. 6): 'a geographically distinct, coherent, functional unit of sub-systems of surface water, soil water and groundwater, subaquatic soils, shores and technical infrastructures for water, including the biotic communities and all associated natural and artificial physical, chemical and biological characteristics and processes.' (Author's note: The boundaries of such systems closely coincide with those of the associated groundwater flow systems.)

Hydrology: the study of the processes and amounts of water exchange between the atmosphere, hydrosphere, and lithosphere.

Hydrosphere: all waters of the Earth in liquid, vapourous and solid state taken together as an entity.

Hydrostatic: qualifier of various properties or parameters of water at rest, e.g., kinetic state, pressure, hydraulic head, and so on.

Hydrostatic head: height of a vertical column of water at rest referenced to a datum plane.

Hydrostatic pressure: pressure of a vertical column of water at rest in the point of measurement.

Hydrostatic pressure gradient or **normal rate of pressure increase:** change in pressure over a unit length of vertical column of water of specified density at rest.

Hydrostratigraphic unit: a distinguishable body of rock of relatively uniform hydraulic conductivity possibly comprising components of different strata, lithology and structure.

Infiltration zone see **groundwater recharge area**.

Inflow region see **groundwater recharge area**.

Intermediate (groundwater) flow system see **groundwater flow-system – intermediate**.

Interval permeability: permeability multiplied by formation thickness (corresponds to transmissivity when hydraulic conductivity is substituted for permeability; the term is used in petroleum geology and reservoir engineering).

Inverse modelling: obtaining distribution of formation parameters (e.g., porosity, permeability, transmissivity, interval permeability, etc.) from observed distribution of hydraulic heads, commonly by iteration.

Isochrone: line connecting points of equal time as, e.g. equal **travel time.**

Isohypse: line connecting points of equal elevation.

Isoline: line connecting points of equal value.

Isotropic: property of rock having **isotropy.**

Isotropy: the property of having equal hydraulic conductivity (permeability) in different directions.

Lag time: difference in time between an initial change in energy (pore pressure, hydraulic head) in one point of the flow domain and its first arrival (the **onset time** or **breakthrough time**) at an other point.

Laplace equation: differential equation relating the spatial distribution of hydraulic heads inside a flow domain to the (temporally invariant) values of hydraulic heads on the domain's boundaries.

Leakage: hydraulic communication, i.e. flow and/or pressure propagation, between **aquifers** across **aquitards.**

Limit anomaly: the maximum possible change in the fluid potential due to gravity flow at one end of a **rock pod.**

Line sink: sink of linear shape.

Liquefaction: the process of transformation from a solid to a liquid state in cohesionless soils due to increased pore pressure and reduced effective stress. (Liquefaction is common in discharge areas and may be manifest by soft quaky ground, **quicksand**, **soap holes**, **mud volcanoes**, and so on.)

Liquefied: soil mechanical condition caused by liquefaction.

Lithosphere: the solid, rocky crust of the Earth, as distinguished from the **atmosphere** and the **hydrosphere.**

Lithotrophic: site condition enriched in plant nutrients (trophic levels enriched by long flow paths of water through the lithosphere; characteristic of groundwater discharge areas).

Local (groundwater) flow system see **groundwater flow system – local.**

Manifestations of groundwater flow: observable, possibly measurable, natural processes and phenomena due to the physical, chemical, and/or dynamic interaction between moving water in the subsurface and its ambient environment.

Mass flux: mass of fluid passing perpendicularly through a unit cross-sectional area during a unit length of time; **specific volume discharge** × density.

Mesophyte: plant preferring average moisture conditions.

Monochronous: adjective applied to a flow field comprising flow systems of similar ages.

Mud volcano (considered in the present context only): popular term (Canadian prairies) for **soap hole.**

Neutral stress see **pore pressure.**

Nominal pore pressure: pore pressure due to the weight of a column of static water of homogeneous density above a point of measurement.

Non-steady state flow see **transient flow.**

Normal rate of pressure increase see **hydrostatic pressure gradient.**

Onset time see **breakthrough time.**

Outflow region see **groundwater discharge area.**

Overpressure see **superhydrostatic pressure.**

p(d)-**profile** or *p(d)*-**curve** see **pressure-depth profile.**

p(z)-**profile** or *p(z)*-**curve** see **pressure-elevation profile.**

Path length: the distance a water particle travels from a repository to the point of its exit from a saturated flow domain at the water table.

Penetration depth or **effectively impermeable base** (see Section 3.1.3; Zijl, 1999): depth below the water table which is reached by an arbitrarily defined proportion (dependent on the purpose) of a flow system's flux.

Permeability (unit of): volume of flow of water of 1 centipoise viscosity through and normal to the 1 cm^2 side of a porous cube during 1 second of time under a 1 atm/cm pressure gradient. It is a property of the rock and not dependent on the water's viscosity and density, thus on temperature. The unit of measurement is called the 'darcy'.

Phreatic belt: the vertical zone of annual fluctuation of the water table.

Phreatophyte: plant with roots habitually reaching the water table and preferring moist conditions.

Piezometric elevation see **piezometric head**.

Piezometric head: misnomer for **hydraulic head**.

Piping: subsurface erosion by groundwater flow resulting in the formation of narrow conduits, 'pipes'.

Plume: an elongated three-dimensional open and mobile column or band of dispersed particles such as smoke, contaminants or radioactive nuclides.

Pore pressure or **formation pressure** or **neutral stress:** pressure acting on the fluid in the pores or other voids of the rock.

Pore-pressure diffusion see **diffusion**.

Pore-pressure profile: graph of pore pressures measured in a vertical borehole, or in different boreholes beneath a limited surface area, plotted against depth or elevation.

Potentiometric anomaly see **anomaly**.

Potentiometric contour: line of constant fluid-potential or hydraulic-head value.

Potentiometric elevation see **hydraulic head**.

Potentiometric surface: imaginary surface connecting the points to which water rises (potentially or really) from a confined aquifer.

Pressure diffusion see **diffusion**.

Pressure gradient: change in pore pressure over a unit length.

Pressure head: height of a fluid column above the point of measurement.

Pressure-depth profile or *p(d)*-**profile** (**curve**): pore-pressure profile referenced to land surface.

Pressure-elevation profile or *p(z)*-**profile** (**curve**): pore-pressure profile referenced to topographic elevation.

Quasi-stagnant zone see **stagnant zone**.

Quicksand: liquefied sand (see **liquefaction**).

Recharge see **groundwater recharge**.

Recharge Area Concept, RCA: theory proposing that judiciously chosen regional recharge areas of groundwater have specifically favourable attributes as sites for high-level radioactive waste repositories.

Recharge area see **groundwater recharge area**.

Reference pattern: pattern of flow or dynamic parameters calculated for flow domains of given boundaries and homogeneous rock framework.

Regional (groundwater) flow system see **groundwater flow system – regional.**

Relative adjustment of hydraulic head: ratio of the induced head-change actually accomplished at a point of observation to the inducing total change at the point of origin.

Relative anomaly: ratio of the **absolute anomaly** to the **limit anomaly** at one end point of a rock pod.

Representative elementary volume (REV): rock body of minimum size the hydraulic properties of which characterize the flow domain considered to be homogeneous.

Return-flow time: duration of travel of a water particle from a nuclear-waste repository to the water table.

Rock framework: the body of all rocks taken together constituting a flow domain.

Rock pod: a three-dimensionally continuous rock body embedded in a rock matrix of different permeability.

Simple drainage basin: drainage basin with linearly sloping water table.

Simultaneity: the contemporaneous occurrence of different actions or processes at different locations.

Singular point see **stagnant zone**.

Sink: location and/or condition in space where fluid vanishes or is withdrawn, from a flow domain, commonly indicated by the termination of flow lines.

Site: place characterized by the combination of conditions determining the type and quality of the local plant association.

Slough: shallow pond, with aquatic and/or phreatophytic vegetation, possibly dry temporarily and/or seasonally.

Soap hole: liquefied part of the land surface characterized by local weakness of limited areal extent underlain by a viscous admixture of sand, silt, clay, and water, brought about by upward moving groundwater. It can have mounded, hollowed, or flat surface, which may change in time (Canadian Prairie vernacular; see **liquefaction**).

Source: location and/or condition in space where fluid is created in, or added to, a flow domain, commonly indicated by the beginning of flow lines.

Specific storage: property of the rock–water system measured by the volume of water that a unit volume of ideally confined rock yields or absorbs under a unit change of hydraulic head in time due to the system's compressibility.

Specific volume discharge: volume of fluid passing perpendicularly through a unit cross-sectional area during a unit length of time – **flux – flow strength – Darcy velocity.**

Specific weight: the force acting upon a horizontal unit surface by a mass of unit volume and density ρ due to gravity. Numerically the specific weight equals the vertical pressure gradient in an unconfined static fluid of density ρ.

Stagnant zone or **quasi-stagnant zone:** part of the flow field of low or zero hydraulic gradients in regions of and due to the convergence from, or divergence towards, opposite directions of two or three different flow systems. The **singular point** of a stagnant zone is its theoretical centre of zero hydraulic gradient.

Static pressure gradient: change in pore pressure over a unit of vertical length in a column of static water.

Static water: water at rest.

Steady-state flow: flow whose direction and intensity is constant in time.

Storage coefficient or **storativity: specific storage** of an ideally confined water-bearing stratum multiplied by its thickness.

Storativity see **storage coefficient**.

Subhydrostatic pressure or **underpressure:** pressure in water lower than that of a vertical column of static water of similar density in the point of measurement.

Subhydrostatic pressure gradient or **subnormal rate of pressure increase:** pressure gradient in a vertical column of water at a given depth that is less than the hydrostatic pressure gradient of water of the same density obtained by determining the tangent to a vertical pressure-profile at the same depth.

Superhydrostatic pressure or **overpressure:** pressure in water greater than that of a vertical column of static water of similar density in the point of measurement.

Superhydrostatic pressure gradient or **supernormal rate of pressure increase:** pressure gradient in a vertical column of water at a given depth that is greater than the hydrostatic pressure gradient of water of the same density obtained by determining the tangent to a vertical pressure-profile at the same depth.

Susceptibility: the tendency or propensity of being subject to an influence or condition.

Thalweg: valley line: the line connecting the lowest points of a valley, or stream bed.

Transient flow or **non-steady state flow:** flow the direction and/or intensity of which varies in time.

Transient pore pressure: pore pressure that varies in time.

Transmissivity: measure of the ability of an ideally confined water-bearing stratum to allow water flow normal to its cross section; it is expressed as the stratum's hydraulic conductivity multiplied by its thickness.

Travel time: the duration of migration of a water particle along a flow path between two specified point.

Trophic level or **status:** degree of adequacy of nourishment in one segment of the food chain characterized by organisms that all obtain food and energy in similar fashions.

Trophic: of or pertaining to nutrition.

Ubiquity: the property of being present everywhere at the same time, omnipresence.

Unconfined: condition in which the upper boundary of a saturated body of groundwater is at atmospheric pressure.

Underpressure see **subhydrostatic pressure.**

Unit basin: drainage basin with two mirror-symmetrical flanks of homogeneous and isotropic rock framework, bounded on top by linear water tables sloping towards a central thalweg, vertical impermeable planes on the sides, and a horizontal impermeable stratum at the base.

Unit of measurement: a subdivision of specified size regarded as a whole part of any dimension. (It is used to characterize and compare magnitudes and states of material objects and natural conditions, for instance, length: mm; area: m^2; volume: m^3; mass: kg; time: s; velocity: $m\,s^{-1}$; acceleration: $m\,s^{-2}$; density: kgm^{-3}; pressure gradient, specific weight: $kgm^{-2}s^{-2}$; see also **Dimension**).

Unsaturated zone or **vadose zone** or **zone of aeration:** the space between the land surface and the water table.

Vadose zone see **unsaturated zone.**

Vertical pressure gradient: change in pore pressure over a unit of vertical length at a specified point or range of depth.

Vertical pressure profile: cross plot of pressure versus depth.

Water divide or **drainage divide:** elevated area of the land surface from which water runs off in diverging directions.

Water table: theoretical surface in a groundwater body on which the pore-water pressure, p, equals atmospheric pressure, p_0. (It separates the capillary fringe from the zone of saturation and is approximated in nature by the elevation of water surfaces in open wells that penetrate only a short distance into the saturated zone.)

Xerophyte: plant preferring arid conditions.

Zone of aeration see **unsaturated zone**

References

AECL (1994a). *Environmental Impact Statement on the Concept for Disposal of Canada's Nuclear Fuel Waste*. AECL-10711, COG-93-1. Chalk River, Ontario: Atomic Energy of Canada Limited, AECL Research.

AECL (1994b). *Summary of the Environmental Impact Statement on the Concept for Disposal of Canada's Nuclear Fuel Waste*. AECL-10721, COG-93-11. Chalk River, Ontario: Atomic Energy of Canada Limited, AECL Research.

Albinet, M. and Cottez, S. (1969). Utilisation et interprétation des cartes de différences de pression entre nappes superposées (Utilization and interpretation of maps of pressure-difference between superjacent aquifers). *Chronique d'hydrogéologie de BRGM, Paris*, **12**, 43–48.

Anderson, M. P. (2005). Heat as a ground water tracer. *Ground Water*, **43**(6), 951–962.

Astié, H., Bellegard, R. and Bourgeois, M. (1969). Contribution à l'étude des différences pièzométriques entre plusieurs aquifères superposés: application aux nappes du tertiaire de la Gironde (Contribution to the study of potentiometric differences between several superposed aquifers: application to the tertiary aquifers of Gironde). *Chronique d'hydrogéologie de BRGM, Paris*, **12**, 49–59.

Back, W. (1960). Origin of hydrochemical facies of groundwater in the Atlantic coastal plain. *Report I, Geochemical Cycles*, XXI Session, Norden, International Geological Congress, Copenhagen, pp. 87–95.

Back, W. (1966). Hydrochemical facies and groundwater flow patterns in northern part of Atlantic Coastal Plain. *Geological Survey Professional Paper, 498-A*. Washington, DC: United States Government Printing Office.

Back, W., Rosenshein, J. S. and Seaber, P. R., eds (1988). *Hydrogeology. Geology of North America*, vol. O-2. Boulder, CO: Geological Society of America.

Badry, A. (compiler) (1972). *A Legend and Guide for the Preparation and Use of the Alberta Hydrogeological Information and Reconnaissance Map Series*. Edmonton, Alberta: Research Council of Alberta. (Also: *Earth Sciences Report 1972–12*, Edmonton, Alberta: Alberta Geological Survey http://www.ags.gov.ab.ca/publications/ESR/PDF/ESR_1972_12.pdf)

Bair, S. E. (1987). Regional hydrodynamics of the proposed high-level nuclear-waste repository sites in the Texas Panhandle. *Journal of Hydrology*, **92**, 149–172.

Bair, E. S. and O'Donnell, T. P. (1985). Hydrodynamics of aquifers and aquitards at proposed high-level nuclear-waste repository sites, Palo Duro Basin, Texas, USA. In *Proceedings 17th Congress, International Association of Hydrogeologists*, Tucson, Arizona, pp. 596–611.

Bars, Y. E., Borschevskiy, G. A., Drod, I. O. and Ovchinnikov, A. M. (1961). Genetic relationship of oil–gas basins to basins of subsurface waters surrounding them. *Petroleum Geology*, **5**(11), 579–586.

Baskov, E. A. (1987). *The Fundamentals of Paleohydrogeology of Ore Deposits*. Berlin: Springer-Verlag.

Batelaan, O. (2006). *Phreatology – Characterizing Groundwater Recharge and Discharge Using Remote Sensing, GIS, Ecology, Hydrochemistry and Groundwater Modeling*. Ph.D. thesis, VUB-Hydrology 47, Vrije Universiteit Brussel, Belgium.

Batelaan, O., De Smedt, F. and Huybrechts, W. (1996). Een kwelkaart voor het Nete-Demer-en Dijlebekken (A map of groundwater up-welling for the Nete, Demer, and Dijle drainage basin). *Water*, **91**, 283–288.

Batelaan, O., and De Smedt, F. D. (2004). SEEPAGE, a new MODFLOW DRAIN Package. *Ground Water*, **42**(4), 576–588.

Batelaan, O. and De Smedt, F. D. (2007). GIS-based recharge estimation by coupling surface–subsurface water balances. *Journal of Hydrology*, **337**(3–4), 337–355, doi:10.1016/j.jhydrol.2007.02.001.

Batelaan, O., De Smedt, F. D. and Triest, L. (2003). Regional groundwater discharge: phreatophyte mapping, groundwater modelling and impact analysis of land-use change. *Journal of Hydrology*, **275**, 86–108.

Beck, A. E., Garven, G. and Stegena, L., eds (1989). Hydrogeological regimes and their subsurface thermal effects. *Geophysical Monograph* 47/IUGG series, vol. 2. Washington, DC: American Geophysical Union.

Bedinger, M. S. (1967). An electrical analog study of the geometry of limestone solution. *Groundwater*, **5**(1), 24–28.

Berry, F. A. F. (1969). Relative factors influencing membrane filtration effects in geologic environments. *Chemical Geology*, **4**, 295–301.

Berry, F. A. F. (1973). High fluid potentials in California coast ranges and their tectonic significance. *American Association of Petroleum Geologists Bulletin*, **57**(7), 1219–1249.

Besbes, M., de Marsily, G. and Plaud, M. (1976). Bilan des eaux souterraines dans le basin Aquitain. In *Hydrogeology of Great Sedimentary Basins*, ed. Rónai, A., *Memoires* 11, International Association of Hydrogeologists. Budapest, Hungarian Geological Institute, pp. 294–303.

Bethke, C. M. (1985). A numerical model of compaction-driven groundwater flow and heat transfer and its application to the paleohydrology of intracratonic sedimentary basins. *Journal of Geophysical Research*, **90**(**B7**), 6817–6828.

Bethke, C. M. (1989). Modeling subsurface flow in sedimentary basins. *Sonderdruck aus Geologische Rundschau*, **78**(1), 129–154.

Bodmer, Ph. (1982). Beitrage zur Geothermie der Schweiz (Contribution to the geothermal conditions of Switzerland). Unpublished Ph.D. thesis, No. 7034, ETH Eidgenossische Technische Hochschule, Zürich, Switzerland.

Bodmer, Ph. and Rybach, L. (1984). *Geothermal Map of Switzerland: Heat Flow Density*. Zürich: Swiss Geophysical Commission.

Bodmer. Ph, and Rybach, L. (1985). Heat flow maps and deep ground water circulation: examples from Switzerland. *Journal of Geodynamics*, **4**(1–4), 233–245.

Boelter, D. H. and Verry, E. S. (1977). *Peatland and Water in the Northern Lake States*. US Department of Agriculture Forest Service, General Technical Report, NC-31.

Bonzanigo, L., Eberhardt, E. and Loew, S. (2007). Long-term investigation of a deep-seated creeping landslide in crystalline rock. Part I. Geological and hydromechanical factors controlling the Campo Vallemaggia landslide. *Canadian Geotechnical Journal*, **44**(10), 1157–1180.

Boyd, S. D. and Murphy, P. J. (1984). *Origin of the Salado, Seven Rivers, and San Andres Salt Margins in Texas and New Mexico*. ONWI/SUB/84/E512-05000.T27, prepared by Stone and Webster Engineering Corporation. Columbus, OH: Office of Nuclear Waste Isolation, Battelle Memorial Institute.

Brace, W. F. (1980). Permeability of crystalline and argillaceous rocks. *International Journal of Rock Mechanics and Mining Sciences*, **17**, 241–251.

Bredehoeft, J. D. and Hanshaw, B. B. (1968). On the maintenance of anomalous fluid pressures. 2. Source layer at depth. *Geological Society of America Bulletin*, **79**, 1107–1122.

Butler, A. P. (1970). Ground water as related to the origin and search for uranium deposits in sandstone. Wyoming Uranium Issue. *Contribution to Geology*, **8**(2) part 1, 81–86.

Cardenas, B. (2007). Potential contribution of topography-driven regional groundwater flow to fractal stream chemistry: residence time distribution analysis of Tóth flow. *Geophysical Research Letters*, **34**, L05403, doi:10.1029/2006GL029126.

Chamberlain, T. C. (1885). The requisite and qualifying conditions of artesian wells. *US Geological Survey Annual Report*, **5**, 125–173.

Chebotarev, I. I. (1955). Metamorphism of natural water in the crust of weathering. *Geochimica et Cosmochimica Acta*, **8**, 22–48, 137–170, 192–212.

Cherry, J., van Everdingen, R. O., Meneley, W. A. and Tóth, J. (1972). *Hydrogeology of the Rocky Mountains and the Interior Plains. Excursion A26. Guidebook*, *XXIV International Geologic Congress*, Montreal, Canada.

Chiarelli, A. (1973). Études des nappes aquifères profondes: contribution de l'hydrogèologie à la connaissance d'un bassin sédimentaire et à l'exploration pétrolière (Study of deep aquifers: hydrogeological contribution to the evaluation of a sedimentary basin and to the exploration for petroleum). Unpublished D.Sc. thesis no. 401, Université de Bordeaux I, Bordeaux (in French).

Clissold, R. J. (1967). Mapping of naturally occurring surficial phenomena to determine groundwater conditions in two areas near Red Deer, Alberta. Unpublished M.Sc. thesis, University of Alberta, Edmonton, Alberta.

Corbet, J. and Bethke, C. M. (1992). Disequilibrium fluid pressures and groundwater flow in the Western Canadian Sedimentary Basin. *Journal of Geophysical Research* **97**(B5), 7203–7217.

D'Andrimont, R. (1906). *La Science Hydrologique*. Revue Universelle des Mines. Liège: Béranges.

Darcy, H. (1856). *Les Fontaines Publiques de La Ville De Dijon*. Paris: Victor Dalmont.

de Vries, J. (1974). Groundwater flow systems and stream nets in the Netherlands. Vrije Universiteit Ph.D. thesis, Amsterdam, Rodopi N.V.

Deere, D. U. and Patton, F. D. (1971). Slope stability in residual soils. In *Proceedings of the Fourth Panamerican Conference on Soil Mechanics and Foundation Engineering*, San Juan, pp. 87–170.

Deming, D. (2002). *Introduction to Hydrogeology*. New York: McGraw-Hill.

Deming, D., Sass, J. H, Lachenbruch, A. H. and De Rito, R. F. (1992). Heat flow and subsurface temperature as evidence for basin scale groundwater flow, North Slope of Alaska. *Geological Society of America Bulletin*, **104**, 528–542.

Deutscher Verband für Wasserwirtschaft und Kulturbau e.V. (1987). *Erkundung Tiefer Grundwasser-Zirkulationssysteme: Grundlagen und Beispiele (Investigation of Deep Groundwater Flow Systems: Principles and Examples)*. Schrifterseihe Deutscher Verband für Wasserwirtschaft und Kulturbau e.V. 81. Hamburg: Parey, p. 223.

Domenico, P. A. (1972). *Concepts and Models in Hydrogeology.* New York: McGraw-Hill.

Domenico, P. A. and Palciauskas V. V. (1973). Theoretical analysis of forced convective heat transfer in regional groundwater flow. *Geological Society of America Bulletin,* **84**, 3803–3814.

Domenico, P. A. and Schwartz, F. W. (1997). *Physical and Chemical Hydrogeology,* 2nd edn. New York: Wiley.

Dupuit, J. (1863). *Études Théoriques et Pratiques sur le Mouvement des Eaux dans les Canaux Découverts et à Travers les Terrains Perméables (Theoretical and Practical Studies of Water Mouvement in Open Channels and Permeable Rocks).* Paris: Dunod.

Eberhardt, E., Bonzanigo, L. and Loew, S. (2007). Long-term investigation of a deep-seated creeping landslide in crystalline rock. Part 2. Mitigation measures and numerical modeling of deep drainage at Campo Vallemaggia. *Canadian Geotechnical Journal,* **44**(10), 1181–1199.

Engelen, G. B. and Jones, G. P., eds (1986). *Developments in the Analysis of Groundwater Flow Systems.* IAHS publication 163. Wallingford: International Association of Hydrological Sciences.

Engelen, G. B. and Kloosterman, F. H. (1996). *Hydrological Systems Analysis: Methods and Applications.* Dordrecht: Kluwer.

Engineering Enterprises, Inc. (1974). Ground-water flow evaluation of areas IX, XIII, and XIV, Red River chloride control project, Oklahoma and Texas, DACW-56-74-C-0088, prepared for United States Army Corps of Engineers, Tulsa District.

England, L. A. and Freeze, R. A. (1988). Finite-element simulation of long-term transient regional groundwater flow. *Ground Water,* **26**(3), 298–308.

Erdélyi, M. (1976). Outlines of the hydrodynamics and hydrochemistry of the Pannonian Basin. *Acta Geologica Academiae Scientiarum Hungaricae,* **20**, 287–309.

Farvolden, R. N. (1961). *A Farm Water Supply from Quicksand.* Preliminary Report **61**(3), Research Council of Alberta, Edmonton, Alberta, Canada.

Fetter C. W. (1994). *Applied Hydrogeology.* Toronto: Maxwell Macmillan Canada.

Fitts, C. R. (1991). Modeling three-dimensional flow about ellipsoidal inhomogeneities with application to flow to a gravel-packed well and flow through lens-shaped inhomogenities. *Water Resource Research,* **27**(5), 815–824.

Fogg, G. E. and Kreitler, Ch. W. (1982). *Groundwater Hydraulics and Hydrochemical Facies in Eocene Aquifers of the East Texas Basin.* Report of Investigation No. 127, Bureau of Econic Geology. Austin, TX: University of Texas, p. 75.

Fourmarier, P. (1939). *Hydrogéologie. Introduction à L'étude des Eaux Destinées à L'Alimentation Humaine et à L'Industrie (Hydrogeology. Introduction to the Study of Water used for Human Consumption and by Industry).* Paris: Masson.

Freeze, R. A. (1969). *Regional Groundwater Flow - Old Wives Lake Drainage Basin, Saskatchewan. Scientific Series* No. 5. Ottawa: Inland Waters Branch, Dept of Energy, Mines and Resources.

Freeze R. A. (1971). Three-dimensional, transient, saturated-unsaturated flow in groundwater basin. *Water Resources Research,* **7**(2), 347–366.

Freeze, R. A. and Cherry, J. A. (1979). *Groundwater.* Englewood Cliffs, NJ: Prentice-Hall.

Freeze, R. A. and Harlan, R. L. (1969). Blueprint for a physically based, digitally simulated hydrologic response model. *Journal of Hydrology,* **9**, 237–258.

Freeze, R. A. and Witherspoon , P. A. (1966). Theoretical analysis of regional groundwater flow: 1. Analytical and numerical solutions to the mathematical model. *Water Resources Research,* **2**(4), 641–656.

Freeze, R. A. and Witherspoon, P., A. (1967). Theoretical analysis of regional groundwater flow. 2. Effect of water table configuration and subsurface permeability variation. *Water Resources Research*, **3**(2), 623–634.

Freeze, R. A. and Witherspoon, P., A. (1968). Theoretical analysis of regional ground water flow. 3. Quantitative interpretations. *Water Resources Research*, **4**(3), 581–590.

Fritz, P. (1968). Osservacioni idrologiche sulle acque di Montecatini Terme (Hydrological observations on the thermal springs of Montecatini Terme). *Bollettino di Geofisica Teorica ed Applicata*, **10**(37), 15–27.

Galloway, W. E. (1978). Uranium mineralization in a coastal-plain fluvial aquifer system: Catahoula Formation, Texas. *Economic Geology*, **73**, 1655–1676.

Galloway, W. E. and Hobday, D. K. (1983). *Terrigenous Clastic Depositional Systems – Applications to Petroleum, Coal, and Uranium Exploration.* New York: Springer-Verlag.

Garven, G. (1989). A hydrogeologic model for the formation of the giant oil sands deposits of the Western Canada Sedimentary Basin. *American Journal of Science*, **289**, 105–166.

Garven, G. and Freeze, R. A. (1984). Theoretical analysis of the role of groundwater flow in the genesis of stratabound ore deposits. 1. Mathematical and numerical model; 2. Quantitative results. *American Journal of Science*, **284**, 1085–1124.

Garven, G., Ge, S., Person, M. A. and Sverjensky, D. A. (1993). Genesis of stratabound ore deposits in the mid-continent Basins of North America. *American Journal of Science*, **293**, 497–568.

Garven, G., Appold, M. S., Toptygina, V. I. and Hazlett, T. J. (1999). Hydrogeologic modeling of carbonate-hosted lead–zinc ore genesis. *Hydrogeology Journal*, **7**(1), 108–126.

Geraghty, J. J. (1960). Movement of contaminants through geologic formations. *1990 Activities.* Urbana, IL: National Water Well Association, Technical Division, pp. 33–43.

Gerrard, A. J. (1981). *Soils and Landforms: an Integration of Geomorphology and Pedology.* London: George Allen and Unwin.

Glaser, P. and Siegel, D. (2007). How Tóthian concepts influenced the modern understanding of peatland hydrology. In *Session* No. 67: *Groundwater Flow I: In Honor of József Tóth, Geological Society of America Annual Meeting*, Denver, Co. Abstracts with programs, vol. **39**, p. 186.

Gleeson, T. and Manning, A. H. (2007). Extending Tóth's examination of regional and local ground-water flow to rugged 3-D mountainous terrain. In *Session* No. 67: *Groundwater Flow I: In Honor of József Tóth, Geological Society of America Annual Meeting*, Denver, CO. Abstracts with programs, vol. **39**, p. 186.

Goff, J. C., Williams, B. P. J., eds (1987). *Fluid Flow in Sedimentary Basins and Aquifers. Geological Society Special Publication*, No. 34. Oxford: Blackwell Scientific.

Gretener, P. E. (1981). *Pore Pressure: Fundamentals, General Ramifications, and Implications for Structural Geology* (Revised). *AAPG Education Course Note Series* No. 4, 2nd edn. Tulsa, OK: American Association of Petroleum Geologists.

Gustafsson, Y. (1970). Topografins inverkan på grundvattenbildningen. Grundvatten (Effect of topography on formation of groundwater. Groundwater). In *Groundwater Problems*, ed. Eriksson, E., Gustaffson, Y. and Nilsson, K. *Proceedings International Symposium.* Stockholm: Pergamon Press, pp. 15–33.

Gustavson, T. C., Finley, R. J. and McGillis, K. A. (1980). *Regional Dissolution of Permian Salt in the Anadarko, Dalhart, and Palo Duro Basins of the Texas*

Panhandle. Report of Investigation No. 106. Austin, TX: Texas Bureau of Economic Geology.

Hagmaier, J. L. (1971). The relation of uranium occurrences to groundwater flow systems. *WGA Earth Science Bulletin*, June, 19–24.

Hanor, J. S. (1987). *Origin and Migration of Subsurface Sedimentary Brines. SEPM (Society for Sedimentary Geology) Lecture notes for short course*, No 21. Tulsa, OK: Society for Sedimentry Geology.

Hantush, M. S. (1956). Analysis of data from pumping tests in leaky aquifers. *Transactions, American Geophysical Union*, **37**, 702–714.

Hantush, M. S. and Jacob, C. E. (1954). Plane potential flow of ground water with linear leakage. *Transactions, American Geophysical Union*, **35**, 917–936.

Hantush, M. S. and Jacob, C. E. (1955). Nonsteady radial flow in an infinite leaky aquifer. *Transactions, American Geophysical Union*, **36**, 95–100.

Harbough, A. W. and McDonald, M. G. (1966). *User's Documentation for MODFLOW-96, an Update to the US Geological Survey Modular Finite-Difference Ground-Water Flow Model*. Open-File Report 96–485. Reston, VA: US Geological Survey.

Harrison, W. J. and Summa, L. L. (1991). Paleohydrology of the Gulf of Mexico Basin. *American Journal of Science*, **291**, 109–176.

Hedberg, H. D. (1996). Foreword. In *Hydrocarbon Migration and its Near-surface Expression*, ed. Schumacher, D. and Abrams, M.A. Memoir vol. 66. Tulsa, OK: American Association of Petroleum Geologists.

Heim, A. (1897). *I movimenti di terreno di Campo V. M. (Of the terrain movements at Campo V. M.)* (unpublished report for the local regional authorities; translated from Germam into Italian by Alberto Totanzi, Cilmamotto, Switzerland).

Heim, A. (1932). *Bergsturz und Menschenleben*. Zürich: Fretz and Wasmuth Verlag (*1989 Landslides and Human Lives*. Vancouver: BiTech Publishers).

Hitchon, B. and Hays., J. (1971). Hydrodynamics and hydrocarbon occurences, Surat Basin, Queensland, Australia. *Water Resources Research*, **7**(3), 650–676.

Hitchon, B., Sauveplane, C. M., Bachu, S., Koster, E. H. and Lytviak, A. (1989). *Hydrogeology* of the Swan Hills Area, Alberta: Evaluation for Deep Waste Injection. *Bulletin* 058. Edmonton, Alberta: Alberta Geological Survey.

Hodgson, G. W. (1980). Origin of petroleum: in-transit conversion of organic components in water. In *Problems of Petroleum Migration*, ed. W H. Roberts, III and R. J. Cordell. *Studies in Geology* No. **10**, Tulsa, OK: American Association of Petroleum Geologists.

Holmes, G. W., Hopkins, D. M. and Foster, H. L. (1968). Pingos in central Alaska. *US Geological Survey Bulletin*, 1241-H.

Holysh, S. and Tóth, J. (1996). Flow of formation waters: likely cause of poor definition of soil gas anomalies over oil fields in east-central Alberta. In *Hydrocarbon Migration and its Near-surface Expression*, ed. Schumacher, D. and Abrams, M. A. *Memoir* 66. Tulsa, OK: American Association of Petroleum Geologists, pp. 255–277.

Hostetler, P. B. and Garrels, R. M. (1962). Transportation and precipitation of uranium and vanadium at low temperatures, with special reference to sandstone-type uranium deposits. *Economic Geology*, **57**(2), 137–167.

HSK (1987). *Technischer Bericht zum Gutachten über das Projekt Gewähr 1985 (Technical Report on Expert Evaluation of Project Guarantee 1985)*. Hauptabteilung für die Sicherheit der Kernanlagen, Bundesamt für Energiewirtschaft, Würenlingen, Switzerland.

Hubbert, M. K. (1940). The theory of ground-water motion. *The Journal of Geology*, **48**: 785–944.

Hubbert, M. K. (1953). Entrapment of petroleum under hydrodynamic conditions. *American Association of Petroleum Geologists Bulletin*, **37**(8), 1954–2026.

Hubbert, M. K. and Rubey, W. W. (1959). Role of fluid pressures in mechanics of overthrust faulting: I. Mechanics of fluid-filled porous solids and its application to overthrust faulting. *Bulletin of the Geological Society of America*, **70**, 115–166.

Hunt, J. M. (1990). Generation and migration of petroleum from abnormally pressured fluid compartments. *American Association of Petroleum Geologists Bulletin*, **74**, 1–12.

Ihrig, D., ed. (1966). *1965 Dunai Árvíz (Danube Flood of 1965). Vízügyi Közlemények, 1966. Évi Külön Kötet (Vízügyi Közlemények, 1966. Special Volume, 1966).* Budapest (in Hungarian with extended abstracts in Russian, English, French and German). Budapest: Vízgazdálkodási Tudományos Kutató Intézet (Research Institute for Water Resources Development).

Ingebritsen, S. E. and Sanford, W. E. (1998). *Groundwater in Geologic Processes.* Cambridge: Cambridge University Press.

Instituto de Geofízico (1985). Logo. *Taller International sobre Acuiferos Regionales (International Workshop on Regional Aquifers).* Mexico City: Universidad Nacional Autónoma de México.

Issar, A. and Rosenthal, E. (1968). The artesian basins of the Central Plateau of Iran. In *Proceedings, International Association of Hydrogeologists, XXIII International Geologic Congress*, Prague. pp. 1–6.

Iverson, R. M. and Reid, M. E. (1992) Gravity-driven groundwater flow and slope failure potential 1. Elastic effective-stress model. *Water Resources Research*, **28**(3), 925–938.

Jacob, C. E. (1940). On the flow of water in elastic artesian aquifers. *Transactions of the American Geophysical Union*, **2**, 574–586.

Jankowski, J. and Jacobson, G. (1989). Hydrochemical evolution of regional groundwaters to playa brines in Central Australia. *Journal of Hydrology*, **108**, 123–173.

Jansson, R., Laudon, H., Johansson, E. and Augspurger, C. (2007). The importance of groundwater discharge for plant species number in riparian zones. *Ecology*, **88**(1), 131–139.

Jones, P. H. and Subramanyam, V. (1961). Ground-water control in the Neyveli lignite field, South Arcot District, Madras State, India. *Economic Geology*, **56**, 273–298.

Jones, G., Fisher, Q. J. and Knipe, R. J., eds (1998). Faulting, fault sealing and fluid flow in hydrocarbon reservoirs. *Geological Society Special Publication* No. 147. London: Geological Society.

Kimmeier, F., Perrochet, P., Andrews, R. and Kiraly, L. (1985). *Simulation par Modèle Mathématique des Ecoulements Souterrains entre Les Alpes et la Forêt Noire; Partie A: Modèle Régional, Partie B: Modèle Local – Nord de la Suisse (Mathematical Modeling of Groundwater Flow between the Alps and the Black Forest; Part A: Regional Model, Part B: Local Model – Northern Switzerland).* Technischer Bericht 84–50. Baden: Nationale Genossenschaft für die Lagerung radioaktiver Abfälle.

Kiraly, L. (1970) L'influence de l'hétérogénéité et de l'anisotropie de la perméabilité sur les systèmes d'écoulement (The influence on flow systems of the heterogeneity and anisotropy of permeability). *Bulletin der Vereinigung Schweizerischer Petroleum-Geologen und -Ingenieure*, **37**(91), 50–57.

Klijn, F. and Witte, J.-P. M. (1999). Eco-hydrology: groundwater flow and site factors in plant ecology. *Hydrogeology Journal*, **7**(1), 65–77.

Kolesov, G. D. (1965). On the question of artesian feeding of rivers. *Soviet Hydrology, Selected Papers*, **3**, 195–203.

Kreitler, C. V. and Bassett, R. L. (1983). Chemical and isotopic composition of saline ground water and saline springs in the Rolling Plains east of the Ogallala escarpment, Texas Panhandle. In *Geology and Geohydrology of the Palo Duro Basin, Texas Panhandle; Report on the Progress of Nuclear Waste Isolation Feasibility Study, 1982.* Geological Circular. Austin, TX: Texas Bureau of Economic Geology, pp. 83–4.

LaFleur, R. G. (1984). *Groundwater as a Geomorphic Agent.* Boston, MA: Allen & Unwin.

Lau, J. E., Commander, D. P. and Jacobson, G. (1987). *Hydrogeology of Australia. Bureau of Mineral Resources, Geology and Geophysics Bulletin* 227. Canberra: Australian Government Publishing Service.

Lazear, G. D. (2006). Evidence for deep groundwater flow and convective heat transport in mountainous terrain, Delta County, Colorado. USA. *Hydrogeology Journal*, **14**(8), 1582–1598.

Leskiw, L. A. (1971). Relationship between soils and groundwater in field mapping near Vegreville, Alberta. Unpublished M.Sc. thesis, University of Alberta, Edmonton, Alberta, Canada.

Mackay, J. R. (1978). Sub-pingo water lenses, Tuktoyaktuk Peninsula, Northwest Territories. *Canadian Journal of Earth Sciences*, **15**(8), 1219–1227.

Macumber, P. G. (1991). *Interaction between Ground Water and Surface Systems in Northern Victoria.* Department of Conservation and Environment, Victoria.

Mádl-Szőnyi, J. (2007). From paradigm to paradigm or from Tóth to Tóth in Hungary's hydrogeology. In *Session* No. 67: *Groundwater Flow I: In Honor of József Tóth. Geological Society of America Annual Meeting*, Denver, Colorado. *Abstracts with Programs*, **39**. Denver, Co: Geological Society of America, p.187.

Margat, J. (1969). Remarques sur la signification des surfaces piézométriques des nappes captives. *Chronique d'Hydrogéologie de BRGM, Paris*, **12**, 13–17.

McDonald, M. G. and Harbough, A. W. (1988). A modular three-dimensional finite-difference ground-water flow model. *Techniques of Water-Resources Investigations*, Book 6, Chapter A. Reston, VA: US Geological Survey.

McGinnis, L. D. and Jensen, T. E. (1971). Permafrost-hydrogeologic regimen in two ice-free valleys, Antarctica, from electrical depth sounding. *Quaternary Research*, **1**(3), 389–409.

Meekes, J. A. C. (1997). High resolution seismic reflection techniques for subsurface models in groundwater hydrology. Ph.D. thesis, Vrije Universiteit, Amsterdam. Delft University Press, Delft.

Meinzer, O. E. (1942). *Hydrology.* New York: Dover.

Meyboom, P. (1962). Patterns of groundwater flow in the Prairie Profile. In *Groundwater. Proceedings of Hydrology Symposium* No.3, November 8–9, Calgary, Alberta. Calgary, Alberta: National Research Council of Canada, pp. 5–20.

Meyboom, P. (1966). Current trends in hydrogeology. *Earth-Science Reviews*, **2**, 345–364.

Meyboom, P., van Everdingen, R. O. and Freeze, R. A. (1966). *Patterns of Groundwater Flow in Seven Discharge Areas in Saskatchewan and Manitoba. Bulletin* 147. Ottawa: Geological Survey of Canada.

Mifflin, M. D. (1968). *Delineation of Ground-Water Flow Systems in Nevada. Publication* No. 4. *Technical Report Series* H.-W. Reno, NV: Center for Water Resources Research, Desert Research Institute, University of Nevada System.

Miller, J. D. and Sias, J. (1998). Deciphering large landslides: linking hydrological, groundwater and slope stability models through GIS. *Hydrological Processes*, **12**, 923–941

Morganti, J. M. (1981). The geology and ore deposits of the Howards stratiform ore deposits: some models and a new classification. *Geoscience Canada*, **8**, 65–75.

Müller, S. W. (1947). *Permafrost or Permanently Frozen Ground and Related Engineering Problems.* Ann Arbor, MI: J. W. Edwards, Inc.

Munn, M. J. (1909). The anticlinal and hydraulic theories of oil and gas accumulation. *Economic Geology*, **4**(6), 509–529.

Nakka, B. W. and Chan, T. (1994). *A Particle-tracking Code (TRACK3D) for Convective Solute Transport Modeling in The Geosphere: Description and User's Manual.* AECL-10881, COG-93-216. AECL Reasearch, Pinawa, Manitoba.

Neumann, S. P. and Witherspoon, P. A. (1971). Transient flow of groundwater to wells in multiple-aquifer systems. In *Aquitards in the Coastal Groundwater Basin of Oxnard Plain, Ventura County. Bulletin* No. 63–4, Appendix A. Department of Water Resources, California, pp. 159–359.

Neumann, S. P. and Witherspoon, P. A. (1972). Field determination of the hydraulic properties of leaky multiple-aquifer systems. *Water Resources Research*, **8**,1284–1298.

Neuzil, C. E., Bredehoeft J. D. and Wolff, R. G. (1984). Leakage and fracture permeability in the Cretaceous shales confining the Dakota Aquifer in South Dakota. In *Proceedings, Theis Conference on Geohydrology. Geohydrology of the Dakota Aquifer*, ed. Jorgensen, D. G. and Signar, D. C. Worthington, OH: National Water Well Association and US Geological Survey, pp. 113–120.

Neuzil, C. E. and Pollock, D. W. (1983). Erosional unloading and fluid pressures in hydraulically "tight" rocks. *Journal of Geology*, **91**(2), 179–193.

Nirei, H. and Furuno, K. (1986). Development of Quaternary resources and environmental protection: status of underground fluid resources use in the Kanto groundwater basin. In *Recent Progress of Quaternary Research in Japan*, ed. Kaizuka, S., Aso, M., Endo, K., *et al.* Tokyo: Science Council of Japan, pp. 71–80.

Obdam, A. N. M. and Veiling, E. J. M. (1987). Elliptical inhomogeneities in groundwater flow. An analytical description. *Journal of Hydrogeology*, **95**, 87–96.

Oliver, J. (1986). Fluids expelled tectonically from orogenic belts: their role in hydrocarbon migration and other geologic phenomena. *Geology*, **14**, 99–102.

Ortega, G. A. and Farvolden, R. N. (1989). Computer analysis of regional groundwater flow and boundary conditions in the basin of Mexico. *Journal of Hydrology*, **110**, 271–294.

Otto, C. (2007). Petroleum hydrogeology–27 years later. In *Session* No. 67: *Groundwater Flow I: In Honor of József Tóth. Geological Society of America Annual Meeting*, Denver, Co. *Abstracts with Programs*, **39**. Denver, Co: Geological Society of America, P.187.

Palmer, Ch. (1911). *The Geochemical Interpretation of Water Analyses. US Geological Survey Bulletin* 479.

Paloc, H. and Back, W., eds (1992). *Hydrogeology of Selected Karst Regions.* International Association of Hydrogeologists, 13. Hannover: Heise.

Parks, K. P. (1989). Groundwater flow, pore pressure anomalies and petroleum entrapment, Belly River Formation, West Central Alberta. Unpublished M.Sc. thesis, Univerisity of Alberta. Department of Geology, Edmonton, Alberta.

Parks, K. P. and Tóth, J. (1995). Field evidence for erosion-induced underpressuring in Upper Cretaceous and Tertiary strata, west central Alberta. *Bulletin of Canadian Petroleum Geology*, **43**(3), 281–292.

Parnell, J. ed., (1994). *Geofluids: Origin, Migration and Evolution of Fluids in Sedimentary Basins. Geological Society Special Publication* No. 78. London: Geological Society.

Pluhowsky, E. J. and Kantrowitz, I. H. (1964). *Hydrology of the Babylon-Islip Area, Suffolk County, Long Island, New York. US Geological Survey Water-Supply* Paper 1768. Washington, DC: US Geological Survey.

Pollock, D. W. (1989) Documentation of computer programs to compute and display pathlines using results from the US Geological Survey modular three-dimensional finite-difference ground-water-flow model. For *US Geological Survey Open-File Report* 89–381. Reston, VI: US Geological Survey.

Pollock, D. W. (1994). *User's Guide for MODPATH/MODPATH-PLOT, Version 3: A Particle Tracking Post-Processing Package for MODFLOW, the US Geological Survey Finite-Difference Ground-Water Flow Model.* Technical Report, Open-File Report 94–464. Reston, VI: US Geological Survey.

Price, R. A. (1994). Cordilleran tectonics and the evolution of the Western Canada Sedimentary Basin. In *The Geological Atlas of the Western Canada Sedimentary Basin*, ed. Mossop, G. D. and Shetsen, I. Calgary: Canadian Society of Petroleum Geologists and Alberta Research Council, pp. 13–24.

Raffensperger, J. P. and Garven, G. (1995a). The formation of unconformity-type uranium deposits: 1. Coupled groundwater flow and heat transport modeling. *American Journal of Science*, **295**, 581–636.

Raffensperger, J. P. and Garven, G. (1995b). The formation of unconformity-type uranium deposits: 2. Coupled hydrochemical modeling. *American Journal of Science*, **295**, 639–696.

Reid, M. E. and Iverson, R. M. (1992). Gravity-driven groundwater flow and slope failure potential 2. Effects of slope morphology, material properties, and hydraulic heterogeneity. *Water Resources Research*, **28**(3), 939–950.

Rich, J. L. (1921). Moving underground water as a primary cause of the migration and accumulation of oil and gas. *Economic Geology*, **16**(6), 347–371.

Romijn, E., Groba, E., Lüttig, G., *et al.*, eds (1985). *Geothermics Thermal – Mineral Waters and Hydrogeology.* Athens: Theophrastus Publications.

Rostron, B. J. and Tóth, J. (1989). Computer simulation of potentiometric anomalies as an aid to exploration for lenticular reservoirs in developed basins. *Geobyte*, August, 39–45.

Rudolf, D. L. and Ferguson, G. (2007). Groundwater flow in an oil sands tailings structure: one Tóthian scale drainage basin and one large environmental legacy. And *Session No. 67: Groundwater Flow I: In Honor of József Tóth. Geological Society of America Annual Meeting*, Denver, Co. *Abstracts with Programs.* Vol. 39. Denver, Co: American Geological Society, p186.

Rybach, L., ed. (1985) *Heat Flow and Geothermal Processes. Proceedings of IUGG Inter-disciplinary Symposium* No. 10, Hamburg, Germany, August 1983. *Journal of Geodynamics*, Special Issue 4.

Sanford, R. F. (1994). A quantitative model of ground-water flow during formation of tabular sandstone uranium deposits. *Economic Geology*, **89**, 341–360.

Sanford, R. F. (1995). Ground-water flow and migration of hydrocarbons to the Lower Permian White Rim sandstone, Tar Sand Triangle, Southeastern Utah. *US Geological Survey Bulletin*, 2000-J.

Sastre Merlin, A. (1978). Hidrogeologia regional de la Cuenca Terciaria del Rio Alberche. Tomo I. Memoria (Regional hydrogeology of the Rio Alberche Tertiary Basin. Volume I). Unpublished Ph.D. thesis, Seccion de Investigation de Recursos Hidraulicos, Instituto Lucas Mallada, Madrid.

Savage, W. Z., Swolfs, H. S. and Powers, P. S. (1985). Gravitational stresses in long symmetric ridges and valleys. *International Journal of Rock Mechanics and Mining Sciences and Geomechanics Abstracts*, **22**(5), 291–302.

Schumacher, D. and Abrams, M. A., eds (1996). *Hydrocarbon Migration and its Near-Surface Expression. Memoir* 66. Tulsa, OK: American Association of Petroleum Geologists.

Schoeller, H. (1935). Utilité de la notion des échange de base pour la comparaison des eaux souterraines (The usefulness of the base-exchange concept for the comparison of groundwaters). *Société Géologie Compte Rendus Sommaire et Bulletin Série,* **5**(5), 651–657.

Schoeller, H. (1955). Géochimie des eaux souterraines. *Revue de l'Institute Français du Pétrol,* **10**, 230–244.

Schoeller, H. (1962). *Les Eaux Souterraines.* Paris: Masson.

Schwartz, F. W. and Domenico, P. A. (1973). Simulation of hydrochemical patterns in regional groundwater flow. *Water Resources Research,* **9**(3), 707–720.

Schwartz, F. W. and Zhang, H. (2003). *Fundamentals of Ground Water.* New York: Wiley.

Senger, R. K., Fogg, G. E. and Kreitler, C. W. (1987). *Effects of Hydrostratigraphy and Basin Development on Hydrodynamics of the Palo Duro Basin, Texas.* Bureau of Economic Geology, Report of Investigations 165. University of Texas, Austin TX.

Seyhan, E., van de Griend, A. A. and Engelen, G. B. (1985). Multivariate analysis and interpretation of the hydrochemistry of a dolomitic reef aquifer, Northern Italy. *Water Resources Research,* **21**(7), 1010–1024.

Shaw, R. D., Shaw, J. F. H., Fricker, H. and Prepas, E. E. (1990). An integrated approach to quantify groundwater transport of phosphorous to Narrow Lake, Alberta. *Limnology and Oceanography,* **35**(4), 870–886.

Shibasaki and Research Group (1995). *Environmental Managements of Groundwater Basins.* Tokyo: Tokai University Press.

SKBF/KBS-3 (1983a). *Final Storage of Spent Nuclear Fuel–KBS-3, I-General.* Swedish Nuclear Fuel Supply Co/Division KBS. Stockholm, Sweden.

SKBF/KBS-3 (1983b). *Final Storage of Spent Nuclear Fuel–KBS-3, III-Barriers.* Swedish Nuclear Fuel Supply Co/Division KBS. Stockholm, Sweden.

Slichter, Ch. S. (1902). The motions of underground waters. *US Geological Survey, Water-Supply and Irrigation Paper,* 67. Washington, DC: US Geological Survey.

Smith, L. and Chapman, D. S. (1983). On the thermal effects of groundwater flow 1. Regional scale systems. *Journal of Geophysical Research,* **88**, 593–608.

Srisuk, K. (1994) Genetic characteristics of the groundwater regime in the Khon Kaen Drainage Basin, Northeast Thailand. Unpublished Ph.D. thesis, University of Alberta, Edmonton, Alberta, Canada.

Stokes, J. and Thunvik, R. (1978). *Investigations of Groundwater Flow in Rock around Repositories for Nuclear Waste.* KBS Teknisk Rapport 47, Inst. för Kulturteknik KTH, Kärn-Bränsle-Säkerhet, Stockholm, Sweden.

Stute, M., Sonntag, C., Deák, J. and Schlosser, P. (1992). Helium in deep circulating groundwater in the Great Hungarian Plain: flow dynamics and crustal and mantel helium fluxes. *Geochimica et Cosmochimica Acta,* **56**, 2051–2067.

Stuyfzand, P. J. (1993). *Hydrochemistry and Hydrology of The Coastal Dune Area of the Western Netherlands.* Nieuwegen: KIVA.

TAC (1992). *Twelfth Annual Report of the Technical Advisory Committee on the Nuclear Fuel Waste Management Program.* Technical Advisory Committee, Atomic Energy of Canada Limited, McMaster University, Hamilton, Ontario.

TAC (1993). *Thirteenth Annual Report of the Technical Advisory Committee on the Nuclear Fuel Waste Management Program.* Technical Advisory Committee, Atomic Energy of Canada Limited, McMaster University, Hamilton, Ontario.

Terzaghi, K. (1925). *Erdbaumechanic auf Bodenphysikalischer Grundlage (Geological Engineering Based on Soil-Mechanical Principles).* Vienna: Franz Deuticke.

Theis, C. V. (1935). The relation between the lowering of the piezometric surface and the rate and duration of discharge of a well using groundwater storage. *Transactions of the American Geophysical Union*, **2**, 519–524.

Thiem, G. (1906). *Hydrogeologische Methoden (Hydrogeological Methods)*. Leipzig: Gebhart.

Thrasher, J., Fleet, A. J., Hay, S. J., Hovland, M. and Düppenbecker, S. (1996). Understanding geology as the key to using seepage in exploration: the spectrum of seepage styles. In *Hydrocarbon Migration and its Near-surface Expression*, ed. Schumacher, D. and Abrams, M. A. *Memoir* 66. Tulsa, OK: American Association of Petroleum Geologists, pp. 279–283.

Tissot, B. P. and Welte, D. H. (1978). *Petroleum Formation and Occurrence*. New York: Springer-Verlag.

Tokarsky, O. (1971). *Hydrogeology of the Grimshaw–Chinook Valley Area, Alberta*. Report 71–2. Research Council of Alberta, Edmonton (One mapsheet); Also: *Hydrogeological Map of the Grimshaw–Chinook Valley Areas, Alberta, NTS 84C/4 and 84C/5*. http://www.ags.gov.ab.ca/publications/ABSTRACTS/MAP_097.html. Edmonton, Alberta: Alberta Geological Survey.

Tóth, J. (1962a). A theory of groundwater motion in small drainage basins in central Alberta, Canada. *Journal of Geophysical Research*, **67**(11), 4375–4387.

Tóth, J. (1962b). A theoretical analysis of groundwater flow in small drainage basins. In *Proceedings of Hydrology Symposium 3, Groundwater*, Calgary 1962. National Research Council of Canada, Associate Committee on Geodesy and Geophysics, Subcommittee on Hydrology, Ottawa. pp. 75–106.

Tóth, J. (1963). A theoretical analysis of groundwater flow in small drainage basins. *Journal of Geophysical Research*, **68**(16), 4795–4812. Reprinted in (1983) *Physical Hydrogeology; Benchmark Papers in Geology*, ed. Freeze, R. A. and Back, W. Stroudsburg, PA: Hutchinson Ross, pp. 328–345.

Tóth, J. (1966a). Mapping and interpretation of field phenomena for groundwater reconnaisance in a prairie environment, Alberta, Canada. *International Association of Scientific Hydrology Bulletin*, **16**(2), 20–68.

Tóth, J. (1966b). *Groundwater Geology, Movement, Chemistry and Resources near Olds, Alberta*. Bulletin 17, Research Council of Alberta. Edmonton, Alberta. Also: Alberta Geological Survey 'Bulletin Series', BUL 017, http://www.ags.gov.ab.ca/publications/ABSTRACTS/BUL_017, Edmonton, Alberta.

Tóth, J. (1968). *A Hydrogeological Study of the Three Hills Area, Alberta*. Bulletin 24, Research Council of Alberta. Edmonton, Alberta, Canada. Also: Alberta Geological Survey 'Bulletin Series', BUL 024, http://www.ags.gov.ab.ca/publications/ABSTRACTS/BUL_024, Edmonton, Alberta.

Tóth, J. (1970). A conceptual model of the groundwater regime and the hydrogeologic environment. *Journal of Hydrology*, **10**(2), 164–176.

Tóth, J. (1971). Groundwater discharge: a common generator of diverse geologic and morphologic phenomena. *International Association of Scientific Hydrology Bulletin*, **16**(1–3), 7–24.

Tóth, J. (1972). Properties and manifestations of regional groundwater movement. *Proceedings, 24th International Geological Congress*, Section 11, Montreal, pp. 153–163.

Tóth, J. (1973). Hydrogeology and yield evaluation of a municipal well field, Alberta, Canada. *Hydrological Sciences Bulletin des Sciences Hydrologiques*, **18**(2), 165–189.

Tóth, J. (1977). The hydrogeological reconnaissance maps of Alberta. In *Alberta Research Council Bulletin*, **35**, 1–11. Also: *Contributions to the Hydrogeology of Alberta*.

Alberta Geological Survey 'Bulletin Series', http://www.ags.gov.ab.ca/publications/ABSTRACTS/BUL_035, Edmonton.

Tóth, J. (1978). Gravity-induced cross-formational flow of formation fluids, Red Earth Region, Alberta, Canada: analysis, patterns and evolution. *Water Resources Research*, **14**(5), 805–843.

Tóth, J. (1979). Patterns of dynamic pressure-increment of formation fluid flow in large drainage basins, exemplified by the Red Earth region, Alberta, Canada. *Bulletin of Canadian Petroleum Geology*, **27**(1), 63–86.

Tóth, J. (1980). Cross-formational gravity-flow of groundwater: a mechanism of the transport and accumulation of petroleum (the generalized hydraulic theory of petroleum migration). In *Problems of Petroleum Migration*, ed. Roberts, W. H. III and Cordell, R. J. Studies in Geology No 10. Tulsa, OK: American Association of Petroleum Geologists, pp. 121–167.

Tóth, J. (1981). Reply – including a discussion of the "tensiometer effect" in cross-formational gravity-flow systems of formation fluids. *Bulletin of Canadian Petroleum Geology*, **29**(1), 135–142.

Tóth, J. (1982). First approximation of groundwater basin-parameters and resources from rudimentary water-level observations in a prairie environment. In *Proceedings, Second National Hydrogeological Conference,* February 4–5, ed. Ozoray, Gy. Winnipeg, Canada, pp. 28–34.

Tóth, J. (1984). The role of regional gravity flow in the chemical and thermal evolution of ground water. In *Proceedings, Practical Applications of Ground Water Geochemistry*, ed. Hitchon, B. and Wallick, E. I. First *Canadian/American Conference on Hydrogeology*. Worthington, OH: National Water Well Association and Alberta Research Council. pp. 3–39.

Tóth, J. (1988). Ground water and hydrocarbon migration. In *Hydrogeology. Geology of North America*, vol. O-2, ed. Back, W. Rosenshein, J. S. and Seaber, P. R. Boulder CO: Geological Society of America, pp. 485–502.

Tóth, J. (1995). Hydraulic continuity in large sedimentary basins. *Hydrogeology Journal*, **3**(4), 4–16.

Tóth, J. (1996) Thoughts of a hydrogeologist on vertical migration and near-surface geochemical exploration for petroleum. In *Hydrocarbon Migration and its Near-surface Expression*, ed. Schumacher, D. and Abrams, M. A. *Memoir* 66. Tulsa, OK: America Association of Petroleum Geologists, pp. 279–283.

Tóth, J. (1999). Groundwater as a geologic agent: an overview of the causes, processes, and manifestations. *Hydrogeology Journal*, **7**(1), 1–14. Also published in Spanish translation: Tóth, J. (2000). Las aguas subterráneas como agente geológico: causas, procesos y manifestaciones. *Boletín Geologico y Minero*, **111**(4), 9–26.

Tóth, J. (2002). József Tóth: an autobiographical sketch. *Ground Water*, **40**(3), 320–324.

Tóth, J. (2003). Hydrogeological reasons for the geotechnical failure of a prairie town's municipal sewage lagoon: a case analysis. In *Proceedings, Annual Conference of the Canadian Geotechnical Society/International Association of Hydrogeologists,* Canadian National Chapter, Winnipeg, Manitoba, Canada, 2003 September 29–October 1. CD-ROM, Paper No. W001.

Tóth, J. (2005). The Canadian School of Hydrogeology: history and legacy. *Ground Water*, **43**(4), 640–644.

Tóth, J. (2007). The story of gravity-driven groundwater flow-systems: in a nutshell. In *Session No. 67: Groundwater Flow I: In Honor of József Tóth, Geological Society of America Annual Meeting*, Denver, CO. *Abstracts with Programs,* vol. 39. Denver, CO: Geological Society of America, p. 187.

Tóth, J. and Almási, I. (2001). Interpretation of observed fluid potential patterns in a deep sedimentary basin under tectonic compression: Hungarian Great Plain, Pannonian Basin. *Geofluids*, **1**(1), 11–36.

Tóth, J. and Corbet, T. (1986). Post-Paleocene evolution of regional groundwater flow-systems and their relation to petroleum accumulations, Taber area, southern Alberta, Canada. *Canadian Petroleum Geology Bulletin*, **34**(3), 339–363.

Tóth, J., Maccagno, M. D., Otto, C. J. and Rostron, B. J. (1991). Generation and migration of petroleum from abnormally pressured fluid compartments: discussion. *American Association of Petroleum Geologists Bulletin*, **75**(2), 331–335.

Tóth, J. and Millar, R. F. (1983). Possible effects of erosional changes of the topographic relief on pore pressures at depth. *Water Resources Research*, **19**(6), 1585–1597.

Tóth, J. and Rakhit, K. (1988). Exploration for reservoir quality rock bodies by mapping and simulation of potentiometric surface anomalies. *Bulletin of Canadian Petroleum Geology*, **36**(4), 362–378.

Tóth, J. and Otto, C. J. (1993). Hydrogeology and oil-deposits at Pechelbronn-Soultz – Upper Rhine Graben. *Acta Geologica Hungarica*, **36**(4), 375–393.

Tóth, J. and Sheng, G. (1996). Enhancing safety of nuclear waste disposal by exploiting regional groundwater flow: the recharge area concept. *Hydrogeology Journal*, **4**(4), 4–25.

Tuan Yung-hou and Chao Hsue-tun. (1968). Hydrogeological features of the Dzungarian artesian basin. *International Geological Review*, **10**(8), 918–933.

Underschultz, J. R., Otto, C. J. and Bartlett, R. (2005). Formation fluids in faulted aquifers: Examples from the Foothills of Western Canada and the North West Shelf of Australia. In *Evaluating Fault and Caprock Seals*, ed. Boult, P. and Kaldi, J. *American Association of Petroleum Geologists*, Hedberg Series, vol. Tulsa, OK: American Association of Petroleum Geologists, pp. 247–260.

Vanden Berg, A. (1969). *Groundwater Chemistry and Hydrology of the Handhills Lake Area, Alberta.* Report 69(1), Research Council of Alberta, Edmonton, Alberta, Canada.

Van der Kamp, G. and Bachu, S. (1989). Use of dimensional analysis in the study of thermal effects of various hydrogeological regimes. In *Hydrogeological Regimes and their Subsurface Thermal Effects*, ed. Beck, A. E. Garven, G. and Stegena, L. *Geophysical Monograph 47/IUGG series*, vol. 2. Washington, DC: American Geophysical Union.

Vereiskii, N. G. and Vostokova, E. A. (1966). *Guidebook for Determining the Lithological Composition of Surface Deposits and Depth Occurrence of Ground Waters.* Jerusalem: Israel Program for Scientific Translations.

Verweij, J. M. (1993). *Hydrocarbon Migration System Analysis. Developments in Petroleum Science*, vol. 35. Amsterdam: Elsevier.

Verweij, J. M. (2003). *Fluid Flow Systems Analysis on Geological Timescales in Onshore and Offshore Netherlands – With Special Reference to the Broad Fourteens Basin.* Utrecht: Netherlands Institute of Applied Geoscience TNO–National Geological Survey.

Vinard, P. H. (1998). *Generation and Evolution of Hydraulic Underpressures in a Marl-Shale Aquitard at Wellenberg, Switzerland.* Ph.D. Thesis. Université de Neuchâtel, Neuchâtel, Switzerland.

Voss, C. I. and Provost, A. M. (2001). *Recharge-Area Nuclear Waste Repository in Southeastern Sweden – Demonstration of Hydrogeologic Siting Concepts and Techniques.* Swedish Nuclear Power Inspectorate (SKI) Report 01:44. Reston, VA: US Geological Survey.

Wallick, E. I. (1981). Chemical evolution of groundwater in a drainage basin of Holocene Age, East Central Alberta, Canada. *Journal of Hydrology*, **54**, 245–283.

Walton, W. C. (1960). *Leaky Artesian Aquifer Conditions in Illinois. Report of Investigations* vol. 39 Urbana, IL: Illinois State Water Survey.

Wang, H. F. and Anderson, M. P. (1982). *Introduction to Groundwater Modeling. Finite Difference and Finite Element Methods*. San Francisco, CA: Freeman.

Wassen, M. J., Barendregt, A., Schot, P. P. and Beltman, B. (1990). Dependency of local mesotrophic fens on a regional groundwater flow system in a poldered river plain in the Netherlands. *Landscape Ecology*, **5**(1), 21–38.

Wells, P. R. A. (1988). Hydrodynamic trapping in the Cretaceous Nahr-Umr Lower Sand of the North area, offshore Qatar. *Journal of Petroleum Technology*. **40**, 357–362.

Wenzel, L. K. (1942). Methods of determining permeability of water-bearing materials. *US Geological Survey Water-Supply Paper* 887. Washington, DC: US Geological Survey.

Williams, R. E. (1968). Groundwater flow systems and related highway pavement failure in cold mountain valleys. *Journal of Hydrology*, **6**(2), 183–193.

Williams, R. E. (1970). Groundwater flow systems and accumulation of evaporate minerals. *American Association of Petroleum Geologists Bulletin*, **54**(7), 1290–1295.

Winter, T. C. (1978). Numerical simulation of steady-state, three-dimensional ground-water flow near lakes. *Water Resources Research*, **14**, 245–254.

Winter, T. C. (2007) The impact of József Tóth on understanding aquatic ecosystem gradients. In *Session* No. 67: *Groundwater Flow I: In Honor of József Tóth, Geological Society of America Annual Meeting*, Denver, CO. *Abstracts with Programs*, vol. 39. Denver, CO: Geological Society of America.

Winter, T. C., Bidwell, L. E. and Maclay, R. W. (1970). *Water Resources of the Wild Rice River Watershed, Northwestern Minnesota*. Hydrologic Investigations, Atlas HA-339. Washington, DC: US Geological Survey.

Winter, T. C., Harvey, J. W., Fanke, O. L. and Alley W. M.(1998). *Ground Water and Surface Water – A Single Resouce. US Geological Survey Circular* 1139. Denver, CO: US Geological Survey.

Witherspoon, P. A., ed. (1989). *Geological Problems in Radioactive Waste Isolation: a World Wide Review. Proceedings of Workshop W3B, 28th International Geological Congress*, Washington, DC, July 15–16. Berkeley, CA: Earth Sciences Division, Lawrence Berkeley Laboratory, University of California.

Witherspoon, P. A. and Neuman, S. P. (1967). Evaluating a slightly permeable caprock in aquifer gas storage: I. Caprock of infinite thickness. *Transactions of the American Institute of Mineral Engineering*, **240**, 949–955.

Wörman, A., Packman, A. I., Marklund, L. and Harvey, J. W. (2007). Fractal topography and subsurface water flows from fluvial bedforms to the continental shield. *Geophysical Research Letters*, **34**, L07402, doi:10.1029/2007GL029426.

Yaalon, D. H. (1963). On the origin and accumulation of salts in groundwater and soils in Israel. *Bulletin of Research Council of Israel*, Section G, **11G**(3), 105–131.

Zaruba, Q. and Mencl, V. (1969). *Landslides and Their Control*. Amsterdam: Czechoslovak Academy of Sciences and Elsevier.

Zijl, W. (1999). Scale aspects of groundwater flow and transport systems. *Hydrogeology Journal*, **7**(1), 139–150.

Zijl, W. and Nawalany, M. (1993). *Natural Groundwater Flow*. Boca Raton, FL: Lewis Publishers.

Zinko, U., Dynesius, M., Nilson, C. and Seibert, J. (2006). The role of soil pH in linking groundwater flow and plant species density in boreal forest landscapes. *Ecography*, **29**(4), 515–524.

Appendices

The derivations of Equations (2.2) and (3.2) have been reproduced below for the sake of possible historical interest with slight stylistic modifications from the author's original manuscripts dated 1961 and 1962, respectively. The verbal comments may reflect the freshly learnt English of the immigrant who landed in Canada in 1960.

Appendix A

*Derivation of the equation for fluid-potential distribution in a
small drainage basin of homogeneous, isotropic medium,
linearly sloping water table, and steady-state groundwater flow.
('The Unit Basin')*

J. Tóth
*Groundwater Division, Research Council of Alberta
Edmonton, Alberta, Canada*

Paper published as:
J. Tóth (1962). A theory of groundwater motion in small drainage basins in central
Alberta, Canada. *Journal of Geophysical Research,* **67**(11), 4375–4387.

The problem is a *boundary-value* problem which can be solved by the Laplace Equation
and appropriate boundary conditions.
The fluid potential Φ for incompressible fluids (Hubbert, 1940):

$$\phi = g \cdot z + \int_{p_0}^{p} \frac{dp}{\rho} = gz \tag{1}$$

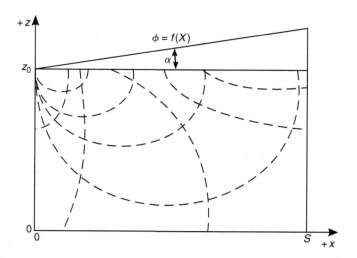

$$\text{Laplace equation:} \frac{\partial^2 \phi}{\partial x^2} + \frac{\partial^2 \phi}{\partial z^2} = 0 \tag{2}$$

275

Boundary conditions:

$$\frac{\partial \phi}{\partial z} = 0 \qquad z = 0, \qquad\qquad\qquad 0 \le x \le s \qquad (3a)$$

$$\frac{\partial \phi}{\partial x} = 0 \qquad x = 0, \qquad\qquad\qquad 0 \le z \le z_0 \qquad (3b)$$

$$\frac{\partial \phi}{\partial x} = 0 \qquad x = s, \qquad\qquad\qquad 0 \le z \le z_0 \qquad (3c)$$

$$\phi = f(x) \qquad \phi = f(x) = g \cdot z = z_0 + c \cdot x, \qquad 0 \le x \le s \qquad (3d)$$

where $c = \tan \alpha$.

General solution by means of separation of the variables:

$$\phi = e^{-kz}(A \cos kx + B \sin kx) + e^{kz}(M \cos kx + N \sin kx) \qquad (4)$$

A, B, M, N are arbitrary constants.

$$\frac{\partial \phi}{\partial z} = -ke^{-kz}(A \cos kx + B \sin kx) + ke^{kz}(M \cos kx + N \sin kx)$$

$$\frac{\partial \phi}{\partial x} = ke^{-kz}(-A \sin kx + B \cos kx) + ke^{kz}(-M \sin kx + N \cos kx)$$

$$\frac{\partial \phi}{\partial x} = 0 \text{ at } x = 0 \text{ no cosine terms possible in } \frac{\partial \phi}{\partial x} = 0$$

thus $B = N = 0$

$$\frac{\partial \phi}{\partial z} = 0 \text{ at } z = 0 \text{ because } B = N = 0$$

$$- A \cos kx + M \cos kx = 0$$

$$A = M$$

$$\frac{\partial \phi}{\partial x} = 0 \text{ at } x = s$$

$$- e^{-kz}A \sin ks - e^{kz}A \sin ks = 0$$

$$A \sin ks \left(e^{-kz} + e^{kz}\right) = 0$$

$$A \sin ks = 0$$

for a nontrivial solution $A \ne 0$

$$\sin ks = 0$$

$$k = \frac{m\pi}{s} \qquad m = 0, 1, 2, \ldots$$

Hence for each value of m there is a solution.

$$\phi_m = e^{-\frac{m\pi z}{s}} A_m \cos \frac{m\pi x}{s} + e^{\frac{m\pi z}{s}} A_m \cos \frac{m\pi x}{s}$$

$$\phi_m = 2A_m \cos \frac{m\pi x}{s} \cdot \left(\frac{e^{\frac{m\pi z}{s}} + e^{-\frac{m\pi z}{s}}}{2} \right)$$

$$\phi_m = C_m \cos \frac{m\pi x}{s} \cosh \frac{m\pi z}{s}$$

$$\phi = \sum_{m=0}^{\infty} C_m \cos \frac{m\pi x}{s} \cosh \frac{m\pi z}{s}$$

$$\phi = f(x) \text{ at } z = z_0; \quad \phi = f(x) \text{ at } z = z_0 + cx$$

$$f(x) = \sum_{m=0}^{\infty} C_m \cos \frac{m\pi x}{s} \cosh \frac{m\pi z_0}{s}$$

$$C_m \cdot \cosh \frac{m\pi z_0}{s} = \frac{2}{s} \int_0^s f(x) \cos \frac{m\pi x}{s} \, dx$$

$$C_m = \frac{2}{s \cdot \cosh \frac{m\pi z_0}{s}} \int_0^s f(x) \cos \frac{m\pi x}{s} \, dx$$

$$C_0 = \frac{1}{2} C_{m=0} = \frac{1}{s} \int_0^s f(x) \, dx$$

$$\phi = \frac{1}{s} \int_0^s f(x) \, dx + 2 \sum_{m=1}^{m=\infty} \frac{\cos \frac{m\pi x}{s} \cosh \frac{m\pi z}{s}}{s \cosh \frac{m\pi z_0}{s}} \int_0^s f(x) \cos \frac{m\pi x}{s} \, dx$$

To find ϕ first we integrate the two integrals:

$$\phi = f(x) = g \cdot z = g(z_0 + cx)$$

1.　$$\frac{1}{s} \int_0^s f(x) \, dx = \frac{g}{s} \int_0^s (z_0 + cx) \, dx$$

$$= \frac{g}{s} \left\{ [z_0 x]_0^s + \left[\frac{cx^2}{2} \right]_0^s \right\} = \frac{g}{s} \left\{ z_0 s + \frac{cs^2}{2} \right\}$$

2. $\displaystyle\int_0^s f(x)\cos\frac{m\pi x}{s}\,dx = g\int_0^s (z_0 + cx)\cos\frac{m\pi x}{s}\,dx$

$$= g\left\{ y_0 \underbrace{\int_0^s \cos\frac{m\pi x}{s}\,dx}_{\text{I}} + c\underbrace{\int_0^s x\cdot\cos\frac{m\pi x}{s}\,dx}_{\text{II}} \right\}$$

$$\text{I} = z_0\left[\frac{s}{m\pi}\sin\frac{m\pi x}{s}\right]_0^s = z_0\left(\frac{s}{m\pi}\underbrace{\sin\frac{m\pi s}{s}}_{0} - \frac{s}{m\pi}\underbrace{\sin\frac{m\pi\cdot 0}{s}}_{0}\right) = 0$$

$$\text{II} = c\left[\underbrace{\int_0^s x\cos\frac{m\pi x}{s}\,dx = \frac{csx}{m\pi}\sin\frac{m\pi x}{s}}_{0}\right]_0^s - \frac{cs}{m\pi}\int_0^s \sin\frac{m\pi x}{s}\,dx$$

$$= -\frac{cs}{m\pi}\left[\frac{-s}{m\pi}\cos\frac{m\pi x}{s}\right]_0^s \qquad u = x \qquad dv = \cos\frac{m\pi x}{s}\,dx$$

$$\qquad\qquad\qquad\qquad\qquad\qquad du = dx \qquad v = \frac{s}{m\pi}\sin\frac{m\pi x}{s}$$

$$= \left[\frac{cs^2}{m^2\pi^2}\cos\frac{m\pi x}{s}\right]_0^s = \frac{cs^2}{m^2\pi^2}\left[\cos\frac{m\pi s}{s} - \cos\frac{m\pi\cdot 0}{s}\right]$$

$$= \frac{cs^2}{m^2\pi^2}[\cos m\pi - 1]$$

$$m = 1 \qquad\qquad \cos m\pi - 1$$
$$m = 2 \qquad\qquad \cos m\pi = 1$$
$$m = 3 \qquad\qquad \cos m\pi = -1$$

$$\text{for } m = 2n \qquad n = 1,2,3,\ldots \qquad \int f(x)\ldots = 0$$

$$\text{for } m = 2n \qquad n = 1,2,3,\ldots \qquad \int f(x)\ldots = -\frac{2cs^2}{m^2\pi^2}$$

$$\int_0^s f(x)\cos\frac{m\pi x}{s}\,d = \begin{cases} 0 & \text{for } m = 0,2,4,6,\ldots \\ -\dfrac{2gcs^2}{m^2\pi^2} & \text{for } m = 1,3,5,7,\ldots \end{cases}$$

Substituting the evaluated integrals in ϕ, we obtain:

$$\phi = \frac{g}{s}\left(z_0 s + \frac{cs^2}{2}\right) - 4\sum_{m=1}^{m=\infty}\frac{gcs^2}{m^2\pi^2}\cdot\frac{\cos\dfrac{m\pi x}{s}\cdot\cosh\dfrac{m\pi z}{s}}{s\cdot\cosh\dfrac{m\pi z_0}{s}};\quad m=1,3,5,7\ldots$$

$$\phi = \frac{g}{s}\left(z_0 s + \frac{cs^2}{2}\right) - \frac{4gcs}{\pi^2}\sum_{m=1}^{m=\infty}\frac{\cos\dfrac{m\pi x}{s}\cdot\cosh\dfrac{m\pi z}{s}}{s\cdot\cosh\dfrac{m\pi z_0}{s}},\tag{5}$$

which is *the solution*.

The solution must satisfy the boundary conditions:

$$\frac{\partial\phi}{\partial x} = -\frac{4gcs}{\pi^2}\sum\frac{-\dfrac{m\pi}{s}\sin\dfrac{m\pi x}{s}\cosh\dfrac{m\pi z}{s}}{m^2\cosh\dfrac{m\pi z_0}{s}}$$

$$\frac{\partial\phi}{\partial x} = \frac{4gc}{\pi}\sum_{m=1}^{m=\infty}\frac{\sin\dfrac{m\pi x}{s}\cosh\dfrac{m\pi z}{s}}{m\cosh\dfrac{m\pi z_0}{s}}$$

$$\frac{\partial\phi}{\partial z} = -\frac{4gcs}{\pi^2}\sum\frac{\cos\dfrac{m\pi x}{s}\cdot\dfrac{m\pi}{s}\sin h\dfrac{m\pi z}{s}}{m^2\cosh\dfrac{m\pi z_0}{s}}$$

$$\frac{\partial\phi}{\partial z} = -\frac{4gc}{\pi}\sum\frac{\cos\dfrac{m\pi x}{s}\sin h\dfrac{m\pi z}{s}}{m\cosh\dfrac{m\pi z_0}{s}};\quad m=1,3,5,7,\ldots$$

At $x=0$
$$\frac{\partial\phi}{\partial x} = \frac{4gc}{\pi}\sum_{m=1}^{m=\infty}\frac{\sin\overbrace{\dfrac{m\pi\cdot 0}{s}}^{0}\cosh\dfrac{m\pi z}{s}}{m\cosh\dfrac{m\pi z_0}{s}} = 0$$

At $x=s$
$$\frac{\partial\phi}{\partial x} = \frac{4gc}{\pi}\sum_{m=1}^{m=\infty}\frac{\sin\overbrace{\dfrac{m\pi s}{s}}^{0}\cosh\dfrac{m\pi z}{s}}{m\cosh\dfrac{m\pi z_0}{s}} = 0$$

$$\text{At } z = 0 \qquad \frac{\partial \phi}{\partial z} = \frac{4gc}{\pi} \sum_{m=1}^{m=\infty} \frac{\cosh \dfrac{m\pi x}{s} \, \sin h \overbrace{\dfrac{m\pi \cdot 0}{s}}^{0}}{m \cosh \dfrac{m\pi z_0}{s}} = 0$$

$$\text{At } z = z_0 \qquad \phi = \frac{g}{s}\left(z_0 s + \frac{cs^2}{2}\right) - \frac{4gcs}{\pi^2} \sum_{m=1}^{m=\infty} \frac{\cosh \dfrac{m\pi x}{s} \, \cosh \dfrac{m\pi z_0}{s}}{m^2 \cosh \dfrac{m\pi z_0}{s}}$$

$$\phi = \frac{g}{s}\left(z_0 s + \frac{cs^2}{2}\right) - \frac{4gcs}{\pi^2} \sum_{m=1}^{m=\infty} \frac{1}{m^2} \cosh \frac{m\pi x}{s} \qquad m = 1,3,5,7,\ldots$$

The terms under the summation sign:

$$\sum_{m=1}^{m=\infty} \frac{1}{m^2} \cos \frac{m\pi x}{s} = \left(\frac{1}{1^2}\cos\frac{1\pi x}{s} + \frac{1}{3^2}\cos\frac{3\pi x}{s} + \frac{1}{5^2}\cos\frac{5\pi x}{s} + \ldots\right)$$

Looking for the original function of this series we find

$$x = \frac{s}{2} - \frac{4s}{\pi^2}\left(\frac{1}{1^2}\cos\frac{\pi x}{s} + \frac{1}{3^2}\cos\frac{3\pi x}{s} + \ldots\right)$$

$$-\left(x - \frac{s}{2}\right)\frac{\pi^2}{4s} = \left(\frac{1}{1^2}\cos\frac{\pi x}{s} + \frac{1}{3^2}\cos\frac{3\pi x}{s} + \ldots\right)$$

We substitute into ϕ:

$$\phi = \frac{g}{s}\left(z_0 s + \frac{cs^2}{2}\right) + \left(x - \frac{s}{2}\right)\frac{\pi^2}{4s}\cdot\frac{4gcs}{\pi^2}$$

$$\phi = g\cdot z_0 + \frac{gcs}{2} + gcx - \frac{gcs}{2}$$

$$\phi = g\cdot z_0 + gcx = g\left(z_0 + cx\right) \tag{3d}$$

This equals $f(x)$, i.e. Eq. 5, thus the four boundary conditions are satisfied:

$$\phi = \frac{g}{s}\left(z_0 s + \frac{cs^2}{2}\right) - \frac{4gcs}{\pi^2} \sum_{m=1}^{m=\infty} \frac{\cos \dfrac{m\pi x}{s} \, \cosh \dfrac{m\pi z}{s}}{m^2 \cosh \dfrac{m\pi z_0}{s}}; \quad m = 1,3,5,\ldots \tag{5}$$

Appendix B

Derivation of the equation for steady-state fluid-potential field in a small drainage basin of homogenous and isotropic rock framework and sinusoidal water-table relief (a 'complex basin').

J. Tóth

Groundwater Division, Research Council of Alberta
Edmonton, Alberta, Canada

Published:
(a) Proceedings of *Hydrology Symposium No. 3- 'Groundwater'*, National Research Council of Canada, Associate Committee on Geodesy and Geophysics, Subcommittee on Hydrology. 1962, p. 75–96.
(b) *Journal of Geophysical Research,* **68**(16) 4795–4812, 1963 August 15.

Elevation (z_t) of any point of the water table with respect to the horizontal impermeable boundary comprises three segments:

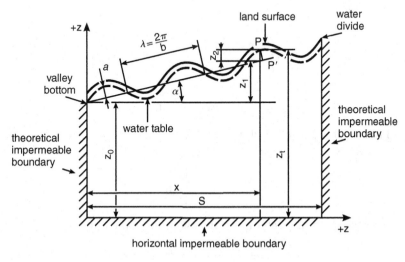

Figure 1. Schematic of idealized domain of ground-water flow in complex basin.

281

(i) $z_0 = $ const.

(ii) $z_1 = x \tan \alpha$

(iii) $z_2 = a \frac{\sin b \left(\frac{x}{\cos \alpha} + p' \right)}{\cos \alpha}$; z_2 may be approximated by neglecting p', if α is sufficiently small.

z_2 then becomes:

$$z_2 = a \frac{\sin \left(b \frac{x}{\cos \alpha} \right)}{\cos \alpha}$$

where $a = $ amplitude $[L]$

$$b = \frac{2\pi}{\lambda} = \text{wave number} \left[\frac{\text{radius}}{L} \right]$$

$\lambda = $ wavelength $[L]$

Thus for any point of the water table:

$$z_t = z_0 + x \tan \alpha + a \frac{\sin \left(b \frac{x}{\cos \alpha} \right)}{\cos \alpha} \qquad (1)$$

For convenience, let:

$$\tan \alpha = c' \left[\frac{L}{L} \right]; \qquad \cos \alpha = c'' \left[\frac{L}{L} \right]$$

$$\frac{a}{c''} = a' \left[\frac{L}{L^2} \right]; \qquad \frac{b}{c''} = b' \left[\frac{\text{radius} \cdot L}{L^2} \right]$$

Then (1) becomes:

$$z_t = z_0 + c'x + \frac{a}{c''} \sin \left(\frac{bx}{c''} \right)$$

and

$$z_t = z_0 + c'x + a' \sin b'x \qquad (1a)$$

The fluid potential is [Eq.16; Hubbert, 1940]:

$$\phi = gz + \int_{p_0}^{p} \frac{dp}{\rho}$$

The integral is zero on the water table, thus Φ reduces there to:

$$\phi = gz_t = g \left(z_0 + c'x + a' \sin b'x \right) \qquad (2)$$

This is a boundary-value problem which can be solved by the Laplace equation (Eq. 1.2). For three of the boundaries the derivatives of ϕ, and for the surface (fourth boundary) ϕ itself can be stated. Thus the four boundary conditions are:

$$\frac{\partial \phi}{\partial x} = 0 \qquad \text{at} \quad x = 0 \quad \text{for} \quad 0 \leq z \leq z_0 \qquad (3a)$$

$$\frac{\partial \phi}{\partial x} = 0 \qquad \text{at} \quad x = s \quad \text{for} \quad 0 \leq z \leq z_0 \qquad (3b)$$

$$\frac{\partial \phi}{\partial z} = 0 \qquad \text{at} \quad z = 0 \quad \text{for} \quad 0 \leq x \leq s \qquad (3c)$$

$$\phi = g\left(z_0 + c'x + a' \sin b'x\right) \quad \text{at} \quad z = z_0 \quad \text{for} \quad 0 \leq x \leq s \qquad (3d)$$

The geometry of the valley (the flow domain) is approximated by a rectangle but the potential distribution is deemed to mimic the water-table relief. This approximation is permissible as long as α is small.

The general solution to the Laplace equation is:

$$\phi = e^{-kz} \left(A \cos kx + B \sin kx\right) + e^{kz} \left(M \cos kx + N \sin kx\right) \qquad (4)$$

where A, B, M, N are arbitrary constants. These constants can be found from the boundary conditions.

Differentiating (Eq. 4) with respect to x and z yields, respectively:

$$\frac{\partial \phi}{\partial x} = ke^{-kz} \left(-A \sin kx + B \cos kx\right) + ke^{kz} \left(-M \sin kx + N \cos kx\right) \qquad (5a)$$

and

$$\frac{\partial \phi}{\partial z} = -ke^{-kz} \left(A \cos kx + B \sin kx\right) + ke^{kz} \left(M \cos kx + N \sin kx\right) \qquad (5b)$$

It is seen from (Eq. 3a) that no cosine terms are possible in (5a), thus:

$$B = N = 0. \qquad (6a)$$

From (Eq. 3c) and (Eq. 5b)

$$-ke^{-kz}A \cos kx + ke^{kz}M \cos kx = \frac{\partial \phi}{\partial z}$$

and

$$-kA \cos kx + kM \cos kx = 0,$$

thus:

$$A = M \qquad (6b)$$

k is found from (3.b):

$$ke^{-kz} \left(-A \sin ks\right) + ke^{kz} \left(-M \sin ks\right) = 0,$$

$$A \sin ks \left(ke^{-kz} + ke^{kz} \right) = 0$$

$$A \sin ks = 0.$$

For a non trivial solution $\sin ks = 0$ and

$$k = \frac{m\pi}{s}; \quad m = 0, 1, 2 \ldots \tag{6c}$$

Using (6.a), (6.b), and (6.c), m particular solutions of (4) are found:

$$\phi_m = e^{-\frac{m\pi}{s}z} A_m \cos \frac{m\pi}{s}x + e^{\frac{m\pi}{s}z} A_m \cos \frac{m\pi}{s}x$$

$$\phi_m = 2A_m \cos \frac{m\pi}{s}x \cdot \frac{e^{\frac{m\pi}{s}z} + e^{-\frac{m\pi}{s}z}}{2}$$

With $2A_m = C_m, \quad \phi_m = C_m \cos \frac{m\pi}{s}x \cdot \cosh \frac{m\pi}{s}z$

The fluid potential is the sum of all particular solutions:

$$\phi = \sum_{m=0}^{\infty} C_m \cos \frac{m\pi}{s}x \cdot \cosh \frac{m\pi}{s}z; \quad m = 0, 1, 2, 3, \ldots \tag{7}$$

Assuming that the surface of the flow domain is a horizontal and straight line, i.e. $z = z_0$, but that the potential varies on that surface according to (3.d.) the coefficient C_m can be found:

$$\text{For } z = z_0 \quad \phi = f(x) = g(z_0 + c'x + a'\sin b'x)$$

Thus from (7)

$$\phi = f(x) = \sum_{m=0}^{\infty} C_m \cos \frac{m\pi x}{s} \cdot \cosh \frac{m\pi}{s}z_0 \tag{8}$$

From Fourier analysis we can write:

$$C_m \cosh \frac{m\pi}{s}z_0 = \frac{2}{s} \int_0^s f(x) \cos \frac{m\pi x}{s} \, dx$$

$$C_m = \frac{2}{s} \cdot \cosh \frac{m\pi z_0}{s} \int_0^s f(x) \cos \frac{m\pi x}{s} \, dx$$

$$C_0 = \frac{1}{2} C_m = \frac{1}{s} \int_0^s f(x) \, dx$$

and from (7) $\phi = C_0 + \sum\limits_{m=1}^{\infty} C_m \cos \dfrac{m\pi x}{s} \cosh \dfrac{m\pi}{s} z$

The potential becomes then:

$$\phi = \frac{1}{s} \int_0^s f(x)\, dx + 2 \sum_{m=1}^{\infty} \frac{\cos \dfrac{m\pi x}{s} \cos \dfrac{m\pi z}{s}}{s \cdot \cosh \dfrac{m\pi z_0}{s}} \int_0^s f(x) \cos \frac{m\pi x}{s}\, dx \qquad (9)$$

where $f(x) = g(z_0 + c'x + a' \sin b'x)$.
Next, the integrals are evaluated:

$$\frac{1}{s} \int_0^s f(x)\, dx = \frac{1}{s} \int_0^s g\left(z_0 + c'x + a' \sin b'x\right) dx = \frac{g}{s} \left[z_0 x + \frac{c'x^2}{2} - \frac{a'}{b'} \cos b'x \right]_0^s$$

$$= \frac{g}{s} \left[z_0 s + \frac{c's^2}{2} - \frac{a'}{b'} \cos b's - \left(0 + 0 - \frac{a'}{b'} \right) \right]$$

$$= \frac{g}{s} \left[z_0 s + \frac{c's^2}{2} + \frac{a'}{b'} \left(1 - \cos b's\right) \right]$$

$$= g \left[z_0 + \frac{c's}{2} + \frac{a'}{sb'} \left(1 - \cos b's\right) \right] \qquad (9a)$$

$$\int_0^s f(x) \cos \frac{m\pi x}{s}\, dx = \int_0^s g\left(z_0 + c'x + a' \sin b'x\right) \cos \frac{m\pi x}{s}\, dx$$

$$= g \left\{ z_0 \underbrace{\int_0^s \cos \frac{m\pi x}{s}\, dx}_{I=0} + c' \underbrace{\int_0^s x \cos \frac{m\pi x}{s}\, dx}_{II} \right.$$

$$\left. + a' \underbrace{\int_0^s \sin b'x \cos \frac{m\pi x}{s}\, dx}_{III} \right\}$$

$$I = z_0 \int_0^s \cos \frac{m\pi x}{s} \, dx = \frac{z_0 s}{m\pi} \int \cos y \, dy = \left[\frac{z_0 s}{m\pi} \sin y \right]_0^{m\pi}$$

$$y = \frac{m\pi x}{s} \qquad dy = \frac{m\pi}{s} \, dx \qquad dx = \frac{s}{m\pi} \, dy$$

$$= \frac{z_0 s}{m\pi} \{ \sin m\pi - \sin 0 \} = 0 \tag{9b.I}$$

$$II = c' \int_0^s x \cos \frac{m\pi x}{s} \, dx = \left[\frac{c' s x}{m\pi} \sin \frac{m\pi x}{s} \right]_0^s - \frac{c' s}{m\pi} \int_0^s \sin \frac{m\pi x}{s} \, dx$$

$$= -\frac{c' s}{m\pi} \left[-\frac{s}{m\pi} \cos \frac{m\pi x}{s} \right]_0^s \qquad u = x \qquad dv = \cos \frac{m\pi x}{s} \, dx$$

$$= -\frac{c' s^2}{m^2 \pi^2} \left[\cos \frac{m\pi x}{s} \right]_0^s \qquad du = dx \qquad v = \frac{s}{m\pi} \sin \frac{m\pi x}{s}$$

$$= \frac{c' s^2}{m^2 \pi^2} \left[\cos \frac{m\pi s}{s} - \cos \frac{m\pi 0}{s} \right] = \frac{c' s^2}{m^2 \pi^2} [\cos m\pi - 1]$$

thus:

for $m = 2n$ where $n = 1, 2, 3, \ldots$ $II = 0$

for $m = 2n - 1$ where $n = 1, 2, 3, \ldots$ $II = -\dfrac{2c' s^2}{m^2 \pi^2}$ \qquad (9b.II)

$$III = a' \int_0^s \sin b' x \cdot \cos \frac{m\pi x}{s} \, dx$$

because

$$\int \sin mx \cdot \cos nx \, dx = -\frac{\cos(m-n)x}{2(m-n)} - \frac{\cos(m+n)x}{2(m+n)}$$

we can write:

$$III = a' \left[-\frac{\cos\left(b' - \dfrac{m\pi}{s}\right)x}{2\left(b' - \dfrac{m\pi}{s}\right)} - \frac{\cos\left(b' + \dfrac{m\pi}{s}\right)x}{2\left(b' + \dfrac{m\pi}{s}\right)} \right]_0^s$$

$$= a' \left\{ \left[-\frac{\cos\left(b' - \frac{m\pi}{s}\right)s}{2\left(b' - \frac{m\pi}{s}\right)} - \frac{\cos\left(b' + \frac{m\pi}{s}\right)s}{2\left(b' + \frac{m\pi}{s}\right)} \right] \right.$$

$$\left. - \left[-\frac{\overbrace{\cos\left(b' - \frac{m\pi}{s}\right)}^{\overbrace{0}^{+1}} \cdot 0}{2\left(b' - \frac{m\pi}{s}\right)} - \frac{\overbrace{\cos\left(b' + \frac{m\pi}{s}\right)}^{\overbrace{0}^{+1}} \cdot 0}{2\left(b' + \frac{m\pi}{s}\right)} \right] \right\}$$

$$= a' \left[-\frac{\cos\left(b's - m\pi\right)}{2\left(b' - \frac{m\pi}{s}\right)} - \frac{\cos\left(b' + \frac{m\pi}{s}\right)s}{2\left(b' + \frac{m\pi}{s}\right)} + \frac{1}{2\left(b' - \frac{m\pi}{s}\right)} + \frac{1}{2\left(b' + \frac{m\pi}{s}\right)} \right]$$

$$= a' \left[\frac{-\left(b' + \frac{m\pi}{s}\right)\cos\left(b's - m\pi\right) - \left(b' - \frac{m\pi}{s}\right)\cos\left(b's + m\pi\right)}{2\left(b'^2 - \frac{m^2\pi^2}{s^2}\right)} \right.$$

$$\left. + \frac{\left(b' + \frac{m\pi}{s}\right) + \left(b' - \frac{m\pi}{s}\right)}{2\left(b'^2 - \frac{m^2\pi^2}{s^2}\right)} \right]$$

$$= a' \left[\frac{\left(b' + \frac{m\pi}{s}\right)\left(1 - \cos\left(b's - m\pi\right)\right) + \left(b' - \frac{m\pi}{s}\right)\left(1 - \cos\left(b's + m\pi\right)\right)}{2\left(b'^2 - \frac{m^2\pi^2}{s^2}\right)} \right]$$

$$= \frac{a'}{2 - \left(b'^2 - \frac{m^2\pi^2}{s^2}\right)} \left\{ \left[b' + \frac{m\pi}{s} \right] \left[1 - \left(\cos b's \cdot \cos m\pi + \sin b's \cdot \overbrace{\sin m\pi}^{0} \right) \right] \right.$$

$$+ \left[b' - \frac{m\pi}{s} \right] \left[1 - \left(\cos b's \cdot \cos m\pi - \sin b's \cdot \overset{0}{\overbrace{\sin m\pi}} \right) \right] \Bigg\}$$

$$= \frac{a'}{2 - \left(b'^2 - \dfrac{m^2\pi^2}{s^2} \right)} \left(b' + \frac{m\pi}{s} - b'\cos b's \cdot \cos m\pi - \frac{m\pi}{s} \cos b's \cdot \cos m\pi \right.$$

$$\left. + b' - \frac{m\pi}{s} - b'\cos b's \cdot \cos m\pi + \frac{m\pi}{s} \cos b's \cos m\pi \right)$$

$$\mathrm{III} = a' \frac{b' - b'\cos b's \cdot \cos m\pi}{b'^2 - \dfrac{m^2\pi^2}{s^2}} \tag{9.b.III}$$

From (9), (9.a), (9.b.I), (9.b.II), and (9.b.III) the final form of the fluid potential is obtained:

$$\phi = g \left\{ z_0 + \frac{c's}{2} + \frac{a'}{sb'}(1 - \cos b's) \right.$$

$$\left. + 2\sum_{m=1}^{\infty} \left[a' \frac{b' - b'\cos b's \cdot \cos m\pi}{b'^2 - \dfrac{m^2\pi^2}{s^2}} - \frac{2c's^2}{m^2\pi^2} \right] \frac{\cos \dfrac{m\pi x}{s} \cosh \dfrac{m\pi z}{s}}{s \cdot \cosh \dfrac{m\pi z_0}{s}} \right\} \quad \text{or}$$

$$\text{where } m = 1, 3, 5, 7, 9, \dots$$

$$\phi = g \left\{ z_0 + \frac{c's}{2} + \frac{a'}{sb'}(1 - \cos b's) \right.$$

$$\left. + \sum_{m=1}^{\infty} \left[\frac{2a'b's}{\pi^2} \cdot \frac{(1 - \cos b's \cdot \cos m\pi)}{\dfrac{s^2 b'^2}{\pi^2} - m^2} - \frac{4\dfrac{c's^2}{\pi^2}}{m^2} \right] \frac{\cos \dfrac{m\pi x}{s} \cosh \dfrac{m\pi z}{s}}{s \cdot \cosh \dfrac{m\pi z_0}{s}} \right\}$$

$$\text{where } m = 1, 3, 5, 7, 9, \dots \quad (10 \text{ and } 3.2)$$

Eq. (10) must satisfy the boundary conditions (3.a), (3.b), (3.c), and (3.d). From (10) we find:

$$\frac{\partial \phi}{\partial x} = -\frac{\dfrac{m\pi}{s} \sin \dfrac{m\pi x}{s} \cosh \dfrac{m\pi z}{s}}{s \cdot \cosh \dfrac{m\pi z_0}{s}} \qquad \begin{array}{l} \text{which is same for both } x = 0 \text{ and } x = s, \\ \text{as required by (3.a) and (3.b).} \end{array}$$

Further on:

$$\frac{\partial \phi}{\partial z} = \frac{\cos \dfrac{m\pi x}{s} \sinh \dfrac{m\pi z}{s}}{s \cdot \cosh \dfrac{m\pi z_0}{s}} = 0 \qquad \text{for } z = 0 \text{ as required by}$$
(3.c).

To check on the fourth boundary conditions (3d), we put $z = z_0$ in (10), and try to find the original functions which, when expanded, form the infinite series. Thus we write the series in separate terms:

(a)
$$2 \sum_{m=1}^{\infty} \frac{a'b'}{s} \cdot \frac{\cos \dfrac{m\pi x}{s}}{b'^2 - \dfrac{m^2\pi^2}{s^2}} = \frac{2a'b'}{s} \sum_{m=1}^{\infty} \frac{\cos \dfrac{m\pi x}{s}}{b'^2 - \dfrac{m^2\pi^2}{s^2}}$$
(11.I)

(b)
$$2 \sum -\frac{a'b' \cos b's \cdot \cos m\pi \cdot \cos \dfrac{m\pi x}{s}}{s \left(b'^2 - \dfrac{m^2\pi^2}{s^2} \right)}$$

$$= -\frac{2a'b' \cos b's}{s} \sum_{m=1}^{\infty} \frac{\cos m\pi \cdot \cos \dfrac{m\pi x}{s}}{b'^2 - \dfrac{m^2\pi^2}{s^2}}$$
(11.II)

(c)
$$-4 \sum_{m=1}^{\infty} \frac{c's^2 \cos \dfrac{m\pi x}{s}}{m^2\pi^2 s} = -\frac{4c's}{\pi^2} \sum_{m=1}^{\infty} \frac{1}{m^2} \cos \frac{m\pi x}{s}$$
(11.III)

Upon arranging the denominator in (11.I), and (11.II), we obtain:

$$b'^2 - \frac{m^2\pi^2}{s^2} = \frac{\pi^2}{s^2} \cdot \frac{s^2}{\pi^2} \left(b'^2 - \frac{m^2\pi^2}{s^2} \right) = \frac{\pi^2}{s^2} \left(\frac{s^2 b'^2}{\pi^2} - m^2 \right)$$

Now we write for (11.II)

$$-\frac{2a'b' \cos b's}{s} \sum_{m=1}^{\infty} \frac{\cos m\pi \cdot \cos \dfrac{m\pi x}{s}}{b'^2 - \dfrac{m^2\pi^2}{s^2}}$$

$$= -\frac{2a'b's\cos b's}{\pi^2} \sum_{m=1}^{\infty} \frac{\cos m\pi \cdot \cos\dfrac{m\pi x}{s}}{\dfrac{s^2 b'^2}{\pi^2} - m^2} \tag{11.II.a}$$

This can be written as:

$$\frac{\cos\alpha z}{\sin\pi z} = \frac{1}{\pi z} + \frac{2z}{\pi} \sum_{m=1}^{\infty} (-1)^n \frac{\cos n\alpha}{z^2 - n^2} \qquad \begin{aligned} &\text{where } n = m, \\ &(-1)^n = \cos m\pi, \\ &\alpha = \tfrac{\pi x}{s}, \\ &z = \tfrac{b's}{\pi}. \end{aligned}$$

↑

(From: An Introduction to the Theory of Functions of a Complex Variable. By: Copson, E.T. Clarendon Press. pp. 448. Oxford.)

On rearranging we find:

$$\sum_{n=1}^{\infty} (-1)^n \frac{\cos n\alpha}{z^2 - n^2} = \frac{\pi}{2z}\left(\frac{\cos\alpha z}{\sin\pi z} - \frac{1}{\pi z}\right) = \frac{\pi}{2\dfrac{b's}{\pi}}\left(\frac{\cos\dfrac{\pi x}{s}\cdot\dfrac{b's}{\pi}}{\sin\pi\dfrac{b's}{\pi}} - \frac{1}{\pi\dfrac{b's}{\pi}}\right)$$

$$= \frac{\pi^2}{2b's}\left(\frac{\cos b'x}{\sin b's} - \frac{1}{b's}\right) \tag{11.II.b}$$

(11.II.b.) is the series in (11.II.a.), so when put back into (11.II.a.) we obtain:

$$-\frac{2a'b'\cdot\cos b's}{\pi^2} \cdot \frac{\pi^2}{2b's}\left(\frac{\cos b'x}{\sin b's} - \frac{1}{b's}\right)$$

$$= -a'\cos b's\left(\frac{\cos b'x}{\sin b's} - \frac{1}{b's}\right) \tag{11.II.c}$$

In looking for the original function of the infinite series in (11.I.) first we convert the product of the two cos series of (11.II.) into one cos function:

$$\cos m\pi \cdot \cos m\frac{\pi x}{s} = \cos\left(m\pi + \frac{\pi x}{s}\right) + \sin m\pi \cdot \sin m\frac{\pi x}{s} = \cos\left[m\left(\frac{\pi x}{s} + \pi\right)\right]$$

When we subtract s from x of this expression we find that the nominator of (11.II) becomes that of (11.I):

$$\cos\left[m\left(\frac{\pi(x-s)}{s} + \pi\right)\right] = \cos\left[m\left(\frac{\pi x}{s} - \frac{\pi s}{s} + \pi\right)\right] = \cos\frac{m\pi x}{s}$$

which proves that $\sum\limits_{m=1}^{\infty} F(x)$ of (11.I) $\sum\limits_{m=1}^{\infty} F(x-s)$ of (11.II). $\qquad$ (11.I.a)

This result is applied to (11.II.b.) to obtain the original function of (11.I):

$$\sum_{m=1}^{\infty} \frac{\cos\left[m\left(\dfrac{\pi(x-s)}{s}+\pi\right)\right]}{\dfrac{s^2 b'^2}{\pi^2}-m^2} = \frac{\pi^2}{2b's}\left(\frac{\cos b'(x-s)}{\sin b's}-\frac{1}{b's}\right)$$

$$= \frac{\pi^2}{2b's}\left(\frac{\cos\left(b'x-b's\right)}{\sin b's}-\frac{1}{b's}\right)$$

$$= \frac{\pi^2}{2b's}\left(\frac{\cos b'x\cdot\cos b's+\sin b'x\cdot\sin b's}{\sin b's}-\frac{1}{b's}\right)$$

$$= \frac{\pi^2}{2b's}\left(\frac{\cos b'x\cdot\cos b's}{\sin b's}+\sin b'x-\frac{1}{b's}\right)$$

$$\text{(11.I.b)}$$

On substituting this back into (11.I.) (taking the proper denominator) we finally obtain the original function:

$$\frac{2a'b's}{\pi^2}\cdot\frac{\pi^2}{2b's}\left(\frac{\cos b'x\cdot\cos b's}{\sin b's}+\sin b'x-\frac{1}{b's}\right)$$

$$= a'\left(\frac{\cos b'x\cdot\cos b's}{\sin b's}+\sin b'x-\frac{1}{b's}\right)\quad\text{(11.I.c.)}$$

Going on to (11.III) we find that the term under the summation sign can be written as an infinite series:

$$\sum_{m=1}^{\infty}\frac{1}{m^2}\cos\frac{m\pi x}{s}=\left(\frac{1}{1^2}\cos\frac{1\pi x}{s}+\frac{1}{3^2}\cos\frac{3\pi x}{s}+\frac{1}{5^2}\cos\frac{5\pi x}{s}+\dots\right)$$

$$\text{(11.III.a)}$$

But $\qquad x = \dfrac{s}{2}-\dfrac{4s}{\pi^2}\left(\dfrac{1}{1^2}\cos\dfrac{1\pi x}{s}+\dfrac{1}{3^2}\cos\dfrac{3\pi x}{s}+\dots\right)$

Thus $\qquad -\left(x-\dfrac{s}{2}\right)\dfrac{\pi^2}{4s}=\left(\dfrac{1}{1^2}\cos\dfrac{\pi x}{s}+\dfrac{1}{3^2}\cos\dfrac{3\pi x}{s}+\dots\right)\qquad\text{(11.III.b)}$

Which, when put back into (11.III) gives:

$$\left(c'x - \frac{c's}{2}\right)$$

(11.III.c)

Eq.(10) is solved for the fourth boundary condition when z_0 is put for z and (11.I.c.), (11.II. c.), and (11.III.c.) are substituted:

$$\phi_{z=z_0} = g \left\{ z_0 + \frac{c's}{2} + \frac{a'}{sb'}\left(1 - \cos b's\right) \right.$$

$$+ a'\left(\frac{\cos b'x \cdot \cos b's}{\sin b's} + \sin b'x - \frac{1}{b's}\right)$$

$$-a'\cos b's\left(\frac{\cos b'x}{\sin b's} - \frac{1}{b's}\right) + \left(xc' - \frac{c's}{2}\right)\right\}$$

$$= g\left\{ z_0 + \frac{c's}{2} + \frac{a'}{sb'} - \frac{a'\cos b's}{sb'} + a'\frac{\cos b'x \cdot \cos b's}{\sin b's} + a'\sin b'x\right.$$

$$\left. - \frac{a}{b's} - \frac{a'\cos b's \cdot \cos b'x}{\sin b's} + \frac{a'\cos b's}{b's} + c'x - \frac{c's}{2}\right\}$$

$$\phi_{z=z_0} = g\left\{z_0 + c'x + a'\sin b'x\right\}$$

(12)

which is the same as (3d). By satisfying the last boundary condition, the correctness of (10) is proven.

Calculation of parameters (constants) for the numerical evaluation of equation 10 (selected examples of those used in Tóth, 1962b and 1963).

Equations and symbols employed in the calculation of the constants:

a = amplitude[ft]; α = regional slope of the water table

λ = wavelength[ft]; $\tan\alpha = c'\left[\dfrac{ft}{ft}\right]$; $\dfrac{a}{c''} = a'\left[\dfrac{ft^2}{ft}\right]$

$b = \dfrac{2\pi}{\lambda}$ = wave number $\left[\dfrac{radians}{ft}\right]$; $\cos\alpha = c''\left[\dfrac{ft}{ft}\right]$; $\dfrac{b}{c''} = b'\left[\dfrac{ft \cdot radians}{ft^2}\right]$

λ and b are kept constants:

$$\lambda = 5000 \text{ ft} \quad b = \frac{2\pi}{\lambda} = 0,00125664\frac{radians}{ft}$$

$\alpha_1 = 0°34'22, 18'';$ $\qquad$ $\alpha_2 = 1°08'44, 75'';$ $\qquad$ $\alpha_3 = 2°51'44, 66''$

$c_1' = \tan \alpha_1 = 0.01 \dfrac{\text{ft}}{\text{ft}};$ $\qquad$ $c_2' = \tan \alpha_2 = 0.02 \dfrac{\text{ft}}{\text{ft}};$ $\qquad$ $c_3' = \tan \alpha_3 = 0.05 \dfrac{\text{ft}}{\text{ft}}$

$c_1'' = \cos \alpha_1 = 0.999950 \dfrac{\text{ft}}{\text{ft}};$ $\quad$ $c_2'' = \cos \alpha_2 = 0.999800 \dfrac{\text{ft}}{\text{ft}};$ $\quad$ $c_3'' = \cos \alpha_3 = 0.998752 \dfrac{\text{ft}}{\text{ft}}$

$a_1 = 50 \text{ ft};$ $\qquad\qquad\qquad$ $a_2 = 100 \text{ ft}$ $\qquad\qquad\qquad$ $a_3 = 200 \text{ ft}$

Problem 1.) $\quad a_{11}' = \dfrac{a_1}{c_1''} = 50.00250012;$ $\quad b_1' = \dfrac{b}{c_1''} = 0.0012567028;$ $\quad c_1' = 0.01$

Problem 2.) $\quad a_{12}' = \dfrac{a_1}{c_2''} = 50.01000200;$ $\quad b_2' = \dfrac{b}{c_2''} = 0.0012568914;$ $\quad c_2' = 0.02$

Problem 3.) $\quad a_{13}' = \dfrac{a_1}{c_3''} = 50.06247797;$ $\quad b_3' = \dfrac{b}{c_3''} = 0.0012582102;$ $\quad c_3' = 0.05$

Problem 4.) $\quad a_{21}' = \dfrac{a_2}{c_1''} = 100.00500024;$ $\quad b_1' = \dfrac{b}{c_1''} = 0.0012567028;$ $\quad c_1' = 0.01$

Problem 5.) $\quad a_{22}' = \dfrac{a_2}{c_2''} = 100.02000400;$ $\quad b_2' = \dfrac{b}{c_2''} = 0.0012568914;$ $\quad c_2' = 0.02$

Problem 6.) $\quad a_{23}' = \dfrac{a_2}{c_3''} = 100.12495594;$ $\quad b_3' = \dfrac{b}{c_3''} = 0.0012582102;$ $\quad c_3' = 0.05$

Problem 7.) $\quad a_{31}' = \dfrac{a_3}{c_1''} = 200.0100048;$ $\quad b_1' = \dfrac{b}{c_1''} = 0.0012567028;$ $\quad c_1' = 0.01$

Problem 8.) $\quad a_{32}' = \dfrac{a_3}{c_2''} = 200.040008;$ $\quad b_2' = \dfrac{b}{c_2''} = 0.0012568914;$ $\quad c_2' = 0.02$

Problem 9.) $\quad a_{33}' = \dfrac{a_3}{c_3''} = 200.24991188;$ $\quad b_3' = \dfrac{b}{c_3''} = 0.0012582102;$ $\quad c_3' = 0.05$

Index

Printed in the United States
By Bookmasters